T0388214

Applications of AI and IOT in Renewable Energy

Applications of AI and IOT in Renewable Energy

Edited by

RABINDRA NATH SHAW
Department of International Relations, Bharath Institute of Higher Education and Research (Deemed to be University), Chennai, India

ANKUSH GHOSH
School of Engineering and Applied Sciences, The Neotia University, Kolkata, India

SAAD MEKHILEF
Swinburne University of Technology, Hawthorn, VIC, Australia

VALENTINA EMILIA BALAS
Department of Automatics and Applied Software, Faculty of Engineering, Aurel Vlaicu University of Arad, Arad, Romania

Academic Press is an imprint of Elsevier
125 London Wall, London EC2Y 5AS, United Kingdom
525 B Street, Suite 1650, San Diego, CA 92101, United States
50 Hampshire Street, 5th Floor, Cambridge, MA 02139, United States
The Boulevard, Langford Lane, Kidlington, Oxford OX5 1GB, United Kingdom

British Library Cataloguing-in-Publication Data
A catalogue record for this book is available from the British Library

Library of Congress Cataloging-in-Publication Data
A catalog record for this book is available from the Library of Congress

ISBN: 978-0-323-91699-8

For Information on all Academic Press publications
visit our website at https://www.elsevier.com/books-and-journals

Publisher: Charlotte Cockle
Acquisitions Editor: Lisa Reading
Editorial Project Manager: Maria Elaine D. Desamero
Production Project Manager: Kamesh Ramajogi
Cover Designer: Miles Hitchen

Typeset by MPS Limited, Chennai, India

Contents

List of contributors

Abdullah Abusorrah
Center of Research Excellence in Renewable Energy and Power Systems, King Abdulaziz University, Jeddah, Saudi Arabia

Ibrahim Alhamrouni
British Malaysian Institute, Universiti Kuala Lumpur, Selangor, Malaysia

Mohamed Dahman Alshehri
Department of Computer Science, College of Computers and Information, Taif University, Taif, Saudi Arabia

Yusuf Alturki
Electrical and Computer Engineering Department, King Abdulaziz University, Jeddah, Saudi Arabia

Kanishk Barhanpurkar
Department of Computer Science, State University of New York, Binghamton, NY, United States

M. Reyasudin Basir Khan
Manipal International University, Negeri Sembilan, Malaysia

Youcef Belkhier
Department of Computer Science, Faculty of Electrical Engineering, Czech Technical University in Prague, Prague, The Czech Republic

Faizah Fayaz
Advance Power Electronics Research Lab, Department of Electrical Engineering, Jamia Millia Islamia (A Central University), New Delhi, India

Ankush Ghosh
School of Engineering and Applied Sciences, The Neotia University, Sarisha, India

Ahteshamul Haque
Advance Power Electronics Research Lab, Department of Electrical Engineering, Jamia Millia Islamia (A Central University), New Delhi, India

Anjali Jain
Department of Electrical and Electronics Engineering, Amity University, Noida, India

Sachin Jain
Senior member IEEE Electrical Engineering Department National Institute of Technology Raipur, Raipur, India

Sephali Shradha Khamari
Department of Electrical Engineering, Indian Institute of Technology, Patna, India

Mohammed Ali Khan
Advance Power Electronics Research Lab, Department of Electrical Engineering, Jamia Millia Islamia (A Central University), New Delhi, India

Pritam Khan
Department of Electrical Engineering, Indian Institute of Technology, Patna, India

Ashwani Kumar
Department of Electrical Engineering, Uttrakhand Technical University, Sudhowala, India

Labh Kumar
Department of Electrical, Electronics and Communication Engineering, Galgotias University, Greater Noida, India

Sudhir Kumar
Department of Electrical Engineering, Indian Institute of Technology, Patna, India

V. S. Bharath Kurukuru
Advance Power Electronics Research Lab, Department of Electrical Engineering, Jamia Millia Islamia (A Central University), New Delhi, India

Azra Malik
Advance Power Electronics Research Lab, Department of Electrical Engineering, Jamia Millia Islamia (A Central University), New Delhi, India

Saad Mekhilef
School of Science, Computing and Engineering Technologies, Swinburne University of Technology, Hawthorn, VIC, Australia; Power Electronics and Renewable Energy Research Laboratory, Department of Electrical Engineering, University of Malaya, Kuala Lumpur, Malaysia; Center of Research Excellence in Renewable Energy and Power Systems, King Abdulaziz University, Jeddah, Saudi Arabia

Vishnu Mohan Mishra
Department of Electrical Engineering, G. B. Pant Engineering College, New Delhi, India

Omair Mohammed
University of the Cumberlands, Williamsburg, KY, United States

Avinash Kumar Pandey
Electrical Engineering Department, National Institute of Technology, Raipur, Raipur, India

Linux Patel
Department of Electrical Engineering, Indian Institute of Technology, Patna, India

Priyaratnam
Department of Electrical and Electronics Engineering, Amity University, Noida, India

Nilesh Kumar Rai
Department of Electrical, Electronics and Communication Engineering, Galgotias University, Greater Noida, India

Anand Singh Rajawat
Department of Computer Science Engineering, Shri Vaishnav Vidyapeeth Vishwavidyalaya, Indore, India

Priyesh Ranjan
Department of Electrical Engineering, Indian Institute of Technology, Patna, India

Rakesh Ranjan
Department of Electrical Engineering, Himgiri Zee University, Dehradun, India

Muhyaddin Rawa
Center of Research Excellence in Renewable Energy and Power Systems, King Abdulaziz University, Jeddah, Saudi Arabia

Saravanan D.
Department of Electrical, Electronics and Communication Engineering, Galgotias University, Greater Noida, India

Rabindra Nath Shaw
Department of International Relations, Bharath Institute of Higher Education and Research (Deemed to be University), Chennai, India

Pradum Shukla
Department of Electrical, Electronics and Communication Engineering, Galgotias University, Greater Noida, India

Brijesh Singh
Department of Electrical and Electronics Engineering, KIET Group of Institutions, Ghaziabad, India

Varsha Singh
Member IEEE Electrical Engineering Department National Institute of Technology, Raipur, Raipur, India

Neelam Verma
Department of Electrical and Electronics Engineering, Amity University, Noida, India

Deepak Yadav
Power Electronics and Renewable Energy Research Laboratory, Department of Electrical Engineering, University of Malaya, Kuala Lumpur, Malaysia

Younes Zahraoui
British Malaysian Institute, Universiti Kuala Lumpur, Selangor, Malaysia

CHAPTER ONE

Machine learning algorithms used for short-term PV solar irradiation and temperature forecasting at microgrid

Younes Zahraoui[1], Ibrahim Alhamrouni[1], Saad Mekhilef[2,3] and M. Reyasudin Basir Khan[4]
[1]British Malaysian Institute, Universiti Kuala Lumpur, Selangor, Malaysia
[2]School of Science, Computing and Engineering Technologies, Swinburne University of Technology, Hawthorn, VIC, Australia
[3]Power Electronics and Renewable Energy Research Laboratory, Department of Electrical Engineering, University of Malaya, Kuala Lumpur, Malaysia
[4]Manipal International University, Negeri Sembilan, Malaysia

1.1 Introduction

The growth of deregulated power grid systems in decentralized power plants has caused issues including the decrease of the power quality and the dangers of rising greenhouse gases (GHGs). The resolution of these issues has resulted in the wide proliferation spread of the microgrid (MG) [1]. The concept of MG brought a novel dimension to the centralized electrical power system. Therefore an MG can essentially provide an optimal conglomeration of distributed energy resources (DERs) and conventional generators [2]. The MGs are incorporating such aspects as artificial intelligence, energy storage systems, and control technologies. Thus this combination makes the MG having unique physical characteristics against cyber-attacks, improve reliability and resilience, and decrease GHG [3].

Forecasting power penetrations from the DERs is essential for input into the energy management system in the MG, especially those renewable energy (RE) resources that are integrated with MG that have an intermittent nature and are variated by meteorological states. Thus proper forecasting of the power dispatch is able to predict the desired amount of power to be delivered by available DERs for power production planning, reliable operation, and improve the economic viability of the MG [4].

Applications of AI and IOT in Renewable Energy.
DOI: https://doi.org/10.1016/B978-0-323-91699-8.00001-2

Forecasting is a function that can be performed in various methods based on several factors such as the amount of data collected and input of the model use. Those factors affect significantly the accuracy of forecasting [5]. In the previous works, one can mainly refer to several techniques that have been applied to predict the power penetration based on forecasting the meteorology such as solar radiation, temperature, and wind speed. Valuable studies used the classical methods such as in Ref. [6], applied model-based deep reinforcement learning model combined with Monte Carlo tree approach to planning the residential MG. In Ref. [7] proposed Markov Transition Matrix to generate the forecasting model to predict the amount of temperature and solar irradiation in the MG using data of 4 years. In Ref. [8] developed an energy management system for scheduling the MG operation using the classic method based on historic meteorological information from the Photovoltaic (PV) plant. In Ref. [9] proposed an intelligent dynamic forecasting algorithm that predicts day-ahead requirement of power from the DERs, providing for load scheduling. Scolari et al. [10] presented an approach to evaluate ultra-short-term and short-term prediction intervals for the solar irradiance using the k-means algorithm. In Ref. [11] the authors introduce the numerical weather prediction (NWP) to schedule a day-head reginal irradiance relies on satellite data observations. In Ref. [12] a Persistence Weather Forecast strategy has been used which held a forecasted temperature and irradiance constant over a predefined forecast period.

Over the years, many researchers were found utilizing machine learning (ML) and deep learning (DL) techniques to build their models having higher performance compared to the classical methods. For example, in Ref. [13] used algorithm-based generalized regression neural network (NN) adapted with a genetic algorithm to improve the local consumption of load for PV power generation MG with plug-in EVs. In Ref. [14] a NN-based NWP model has been developed for MG considering many factors and weather parameters such as cloud coverage and temperature. The proposed work used to predict the generated power in the residential MG. In Ref. [15] the authors proposed long short-term memory (LSTM) and recurrent neural network (RNN) for the time series forecasting, the developed model has eliminated some shortcomings of the previous studies as the exploding gradient during the training. However, some studies applied those techniques to predict the short-term solar irradiation and temperature such as, In Ref. [16] the K-nearest Neighbors (KNNs) is utilized to classify the daily local weather types of day-ahead short-term solar

PV solar irradiation and temperature prediction using the operation data from MG connected with the main grid. In Ref. [17] the authors applied the linear regression (LR) model to forecast solar irradiation. In this model, the intra-day time-changing manner of output power was integrated into the prediction of the short-term framework. Shi et al. [18] a day-ahead solar power generation prediction model of short-term is presented for a PV station based on the weather and historical power output data using the support vector machines (SVMs), Table 1.1 elaborates the advantages and shortcomings of the existing ML and DL techniques used in short-term solar irradiation and temperature forecasting.

This work addresses these issues and contributes to the following topics. Firstly, we have presented the related works. Then, we explained the proposed ML methods for short-term PV solar radiation and temperature forecasting such as LR, RF regression, KNN, SVM models using the feature insertion using the temporal attribute. The next section presents the experimental setup and discusses the results acquired throughout a series of a real dataset from PV station constituted of balanced datasets under different factors, and the ML techniques proposed has been compared. Finally, the conclusion.

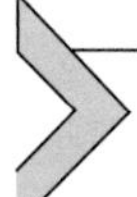

1.2 Proposed work

1.2.1 Overview

The head time period for output prediction of the solar irradiation and temperature is considered the forecast horizon, Therefore the previous literatures classified the forecast horizon into three types based on the amount of the data set as shown in the Table 1.2 [4].

Short-term: is used in the prediction period from second to less than 30 minutes. However, other researchers have considered between 1 minute and more than hours. This type of forecasting is very important to schedule the DERs and minimize the operation cost and restoration process.

Medium-term: can span between 6 and 24 hours, 1 day, 1 week, and a month ahead. It is necessary for maintenance planning, and medium scheduling of the DERs and loads.

Table 1.1 Different machine learning and deep learning techniques applied in short-term forecasting model.

Ref.	Technique	Forecast horizon	Advantages	Disadvantages
Cao and Cao [19]	ANN	Short-term forecasting	• Handle a large number of data set. • Ability to work with incomplete data. • Use the distributed memory	• Requires high hardware performance. • Training data needs time to process.
Pedro and Coimbra [20]	KNN	Short-term forecasting	• Easy to implement. • Does not require training during the prediction process. • Does not learn from the training.	• Scaling the parameters. • Does not work with large data. • Does not work with large dimension data.
Manoj Kumar and Subathra [21]	Random forest (RF)	Short-term forecasting	• Can use to solve classification problem. • Handle the missing values easily. • Less impacted by noise.	• Complexity of implementation. • Scaling the parameters. • Requires much more time to train.
Cao and Lin [22]	RNN	Short-term forecasting	• Process inputs of any length. • Powerfulin time series predictor.	• Slow computation. • Training data can be difficult. • Gradient exploding.
Wang et al. [16]	SVM	Short-term forecasting	• Effectiveness in high dimensional. • The problem of overfitting is less. • Useful in unknowledge dataset.	• Long training time. • Difficulty to tuning the hyper parameters.
Ramsami and Oree [23]	LR	Short-term forecasting	• Easier to implement. • Handles overfitting easily. • More powerful r in the linear data set.	• Sensitive to outliers. • Prone to underfitting.

Table 1.2 Different machine Learning and deep learning techniques have been used in forecasting the solar irradiation and temperature based on the data set amount.

Authors	Country of PV data set	Forecast horizon	Forecast error	Method	Amount of the data set
Zarrindast and Oveisi [24]	China	160 h	Root mean square error (RMSE) = 10%	SVM	The data set collected is 11 months, with a data interval of 15 min, and a daily average solar irradiation time of 13 h.
Moradzadeh et al. [5]	Africa	24 h	RMSE = 3.243%	LSTM	The collected dataset contains 5760 samples and 11 features as input of the model proposed.
Aslam et al. [25]	Korea	24 h	RMSE = 5.6705%	RF	The historical hourly and daily solar irradiation data were collected from 2000 to 2016 and the amount of input features is 11.
Kang et al. [26]	Korea	24 h	RMSE = 10%	LR	The data set has been collected is 4 years, with analyzing solar irradiation data of three time periods 11 a.m, 1 p.m., and 3 p.m.
Kumar et al. [15]	UAE	24 h	RMSE = 11.62%	RNN	The amount of the data set is 744 h started from January.
Leva et al. [27]	Italy	24–72 h	RMSE = 15%	ANN	The analysis data consisted in 240 days of forecasting, with three inputs used in the forecasting model.
Dolara et al. [28]	United States	24 h	RMSE = 10.03	Persistent model	The data was collected from November 2009 to August 2011. Data prior to January 2011 was used as training data.

Long-term: used to predict 24 hours, months, and years in advance. This type of forecasting usually utilizes in seasonal forecasting, sizing the utilities, and forecasting the operation cost in long term.

1.2.2 Different approaches

ML is a systematic process and statistical computation of obtaining the information model from input data without taking an arithmetic operation in rationalized outputs. This systematic process focuses on prediction making using computers and several mathematical optimization equations, which can handle linear, nonlinear, and nonstationary datasets. ML is classified into three types of tasks, namely supervised, unsupervised and semi-supervised, depending on data used [29].

Recently, there has been growing interest in applying the ML in MG to predict the power dispatch from the intermittency DERs, in order to enhance the prediction accuracy and minimize the error. Therefore a significant amount of historical time series data of DERs parameters output and corresponding meteorological variables are used to set up the prediction model of power generation in the MG. The historical series data in the forecasting model are divided into two sets: the training and testing data. The training data are utilized for learning of the model to predict future values, while the testing data are used to validate the forecasting model and calculate the error value of the model considered.

In the section below, we describe the most popular supervised ML techniques that has been used for short-term forecasting for solar irradiation and temperature.

1.2.2.1 Linear regression

LR is the most common predictive model in ML to spot the relation values of the inputs. The mathematical of this supervised can be presented as follows. Given data set as: $D = \left\{ \left(x^{(1)}, y^{(1)}\right), \left(x^{(2)}, y^{(2)}\right), \ldots, \left(x^{(M)}, y^{(M)}\right) \right\}$ with $x^{(m)}, y^{(m)} \in R$ for $m = 1, \ldots, M$. The aim is to predict the values y based on the input values x. The linear model of LR can be given by:

$$f(x, w) = x^T \boldsymbol{w} + \boldsymbol{b} \tag{1.1}$$

where $\boldsymbol{w}, \boldsymbol{b}$ are a constant value in the function $f(x, w)$, and T is the feature vector of inputs x. It should be mentioned that the input $x^{(m)}$ vector needs to include a column of ones. the input x can be represented by a nonlinear function in the space of the data set illustrated, therefore the

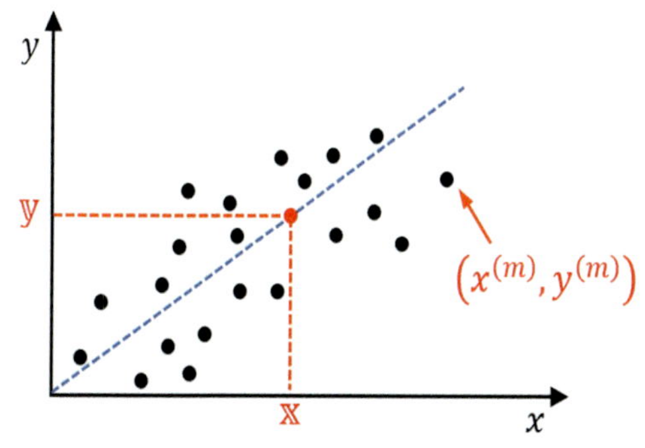

Figure 1.1 Presentation of linear regression using $(\boldsymbol{x}^{(m)}, \boldsymbol{y}^{(m)})$ as training data.

LR is able to proportion the nonlinear objectives as represented in Fig. 1.1.

1.2.2.2 Random forest regression

RF is a regression technique that integrates the efficiency of various ML techniques to forecast the value of a variable. The RF can be utilized for regression-type or classification problems by receiving an (x) input vector and obtain the estimation of the dependent variable. The input data is randomly sampled to generate a forest of regression trees [30]. The RF predictor model is defined by:

$$f(x) = \frac{1}{K}\sum_{k=1}^{k} T(x) \tag{1.2}$$

where K are the individual trees in the forest, and $T(x)$ is the trees grown by $\{T(x)\}_1^K$.

The decision tree is a nonparametric supervised approach that can result in a decision based on the data set with labels and use the function of the tree to prepare an appropriate model to predict future values. At each decision of RT, the mean of the samples and the mean square error should be calculated. In general, to obtain the optimal model with more potential, the LR in the set model must be unrelated as possible [31].

1.2.2.3 K-nearest neighbor

The KNN algorithm is an ML technique for classifying objects based on the closest training examples in the feature space. The simplicity of the KNN concept makes it the preferable model for classification and prediction in different applications [16]. KNN is based on the supposition that

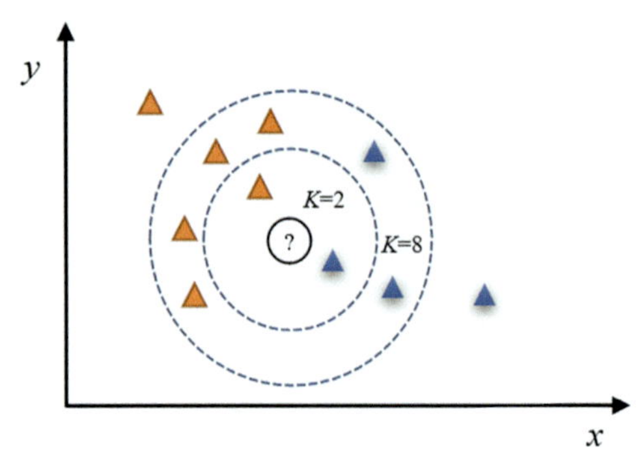

Figure 1.2 The schematic of the K-nearest neighbor classification.

objects of neighbors have related forecasting values. The concept of KNN is described as follows: First of all, find the average value between the new and the training samples, then, according to the classification of the neighbors, obtain the classification of the new sample. the new sample is categorized to the nearest category as illustrated in Fig. 1.2.

However, the new sample is obtained based on rules. This rule is called the "voting KNN rule." Moreover, the categorization does not use any predicting model for fitting and uses the memory to cumulate the feature vectors and class labels of the training samples. The nearest element on the space is obtained according to the standard Euclidean distance law and given by:

$$d_j = \sqrt{\sum\nolimits_{i=1}^{k} (x_i - y_i)^2} \tag{1.3}$$

where x_i, y_i are two elements in the space. k is generated automatically in the model construction process. the forecasting is computed from the time-series values d_j. This approach requires an expanded research space to investigate different values for the parameter of the element and find a value that gives proper results of the model.

1.2.2.4 Support vector machines

The SVM is an effective nonlinear ML method that has been applied as a data mining skill to solve complex problems. SVM is relying on the statistical learning methods and optimization theory. SVM is characterized by its effective performance compared to other models, simple to process the data set and avoid overfitting. Originally, it was developed as a linear classification method, firstly, it was developed to handle a linear classification,

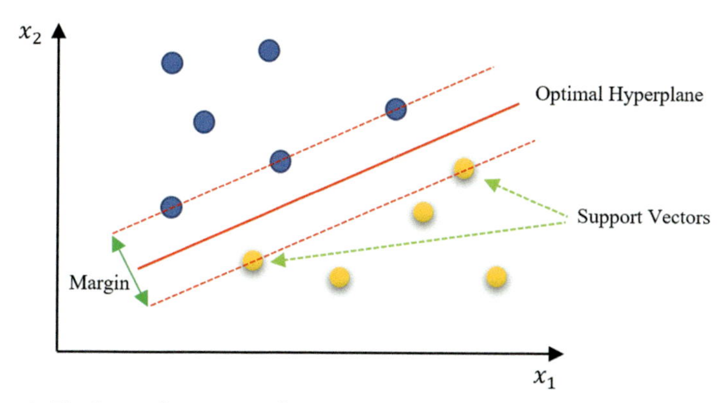

Figure 1.3 The basic illustration of support vector machine classification.

recently, it was generalized to solve the nonlinear problem. The SVM differs from other separating ML methods, using the training set elements to create the hyperplane as shown in Fig. 1.3. The model attempts to obtain the maximum edge to classify between the elements using the hyperplane in the middle. Substantially, the training elements that define the hyperplane are called support vectors.

Mathematically, the objective of the margin in SVM maximize the margin categorization distance $\frac{2}{||\to_w||}$ and reduce the training errors during the training stage and given as [32]:

$$\text{minErro} = \frac{2}{||\to_w||} + C\sum_{i=1}^{m} \delta_i \tag{1.4}$$

where C is the regularization parameter, and δ_i is the training errors. After optimization of Eq. (1.4), the classification decision of a new test data set and create the hyperplane can be determined by solving the equation below:

$$f = \text{sgn}\left(\sum_{i=1}^{m} \alpha_i y_i w + b\right) \tag{1.5}$$

where α_i, y_i are the parameters weight vector, m is the number of support vectors, and b is the bias.

1.2.3 Forecasting accuracy evaluation and validation

Generally, centralized power resources are designed to have a steady output power. However, the load demand fluctuations may result in a vulnerable and unbalance system. Moreover, the fluctuations in the PV plants output are resulting from various factors such as solar irradiation, temperature, and wind speed [33]. In order to evaluate the predicted values of the power dispatch from the PV station, a relation index is calculated which obtain the error variance as given below:

$$R^2 = 1 - \left(\frac{\mathrm{var}(y-1)}{\mathrm{var}(y)}\right) \tag{1.6}$$

The performance of the trained model propose is evaluated based on various metrics using several parameters such as Mean Absolute Error (MAE), RMSE, and Mean Bias Error (MBE). The mathematical equation of these parameters is described in Eqs. (1.1), (1.2), and (1.3), respectively.

$$\mathrm{MAE} = \frac{1}{N}\sum_{t=1}^{N} |y_t - y| \tag{1.7}$$

$$\mathrm{RMSE} = \sqrt{\frac{1}{N}\sum_{t=1}^{N} (y_t - y)^2} \tag{1.8}$$

$$\mathrm{MBE} = \frac{1}{N}\sum_{t=1}^{N} (y_t - y) \tag{1.9}$$

where y_t is the forecasting result at each time point y is the historical data at each time point, and N is the data sample scale, which stands for the time scale in this chapter.

1.2.4 Data creation

Reducing the size dataset can effectively decrease the modeling complexity and ameliorate the computational capability. hence, to create a PV dataset, the interimpact, and relationship between PV and geographic location and other factors should be analyzed first. To find out the base for forecasting, it is substantial to analyze the relation of these data elements. In this research, the dataset was collected from a real PV plant installed at the University of Malaya, Malaysia. There is no centralized to collect the data, therefore most of the collection has been done manually from meteorological devices. Additionally, the illustration of the subsets of data factors used is described below in Fig. 1.4. We select the ideal time slice length for discretizing meteorological devices data dt = 5 minutes.

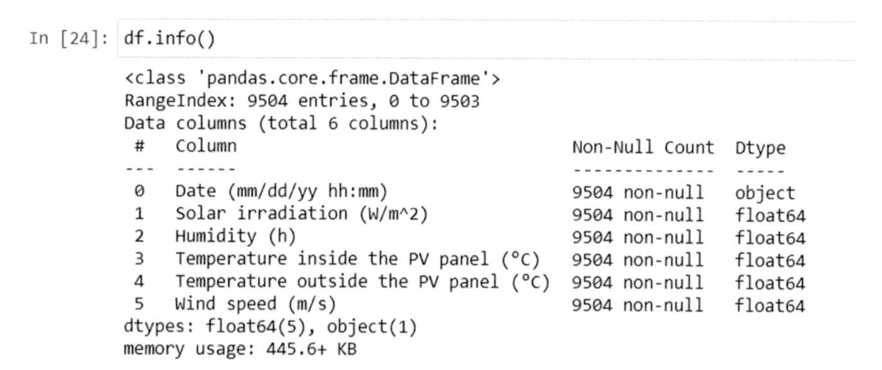

	Date (mm/dd/yy hh:mm)	Solar irradiation (W/m^2)	Humidity (h)	Temperature outside of PV panel (°C)	Temperature outside of PV panel (°C).1	Wind speed (m/s)
0	7/20/2019 0:00	0.13	27815.72	28.39	27.73	0.33
1	7/20/2019 0:05	0.25	27815.81	28.48	27.83	0.18
2	7/20/2019 0:10	0.43	27815.89	28.43	27.81	0.06
3	7/20/2019 0:15	0.87	27815.97	28.56	27.84	0.11
4	7/20/2019 0:20	0.93	27816.05	28.60	27.93	0.37
...	...	...	...	...	...	...
9499	8/21/2019 23:35	0.94	28600.84	25.52	24.93	0.00
9500	8/21/2019 23:40	1.00	28600.92	25.61	24.93	0.00
9501	8/21/2019 23:45	0.94	28601.01	25.63	24.93	0.00
9502	8/21/2019 23:50	1.00	28601.09	25.73	24.96	0.00
9503	8/21/2019 23:55	0.94	28601.17	25.75	25.03	0.00

9504 rows × 6 columns

Figure 1.4 Input data to the model for training and testing.

```
In [24]: df.info()

<class 'pandas.core.frame.DataFrame'>
RangeIndex: 9504 entries, 0 to 9503
Data columns (total 6 columns):
 #   Column                                  Non-Null Count  Dtype
---  ------                                  --------------  -----
 0   Date (mm/dd/yy hh:mm)                   9504 non-null   object
 1   Solar irradiation (W/m^2)               9504 non-null   float64
 2   Humidity (h)                            9504 non-null   float64
 3   Temperature inside the PV panel (°C)    9504 non-null   float64
 4   Temperature outside the PV panel (°C)   9504 non-null   float64
 5   Wind speed (m/s)                        9504 non-null   float64
dtypes: float64(5), object(1)
memory usage: 445.6+ KB
```

Figure 1.5 The information of the data set created.

Analysis of solar irradiation and temperature using ML requires various types of data such as date (Date), solar irradiation (W/m^2), humidity (h), the temperature outside the PV panel (°C), the temperature inside the PV panel (°C), and wind speed (m/s). All these types of factors are essential to training the ML model on what impacts solar irradiation and temperature. Pre-processing further removes all unknown values to avoid null values and the dataset is made precise and normalized. The information of the data set is illustrated in Fig. 1.5.

1.3 Simulation results and comparison

As discussed in the data creation section, historical hourly and daily solar irradiation, and temperature data from July 20, 2019 to August 21, 2019 were used to train and test data, while August 22, 2019 hourly is predicted.

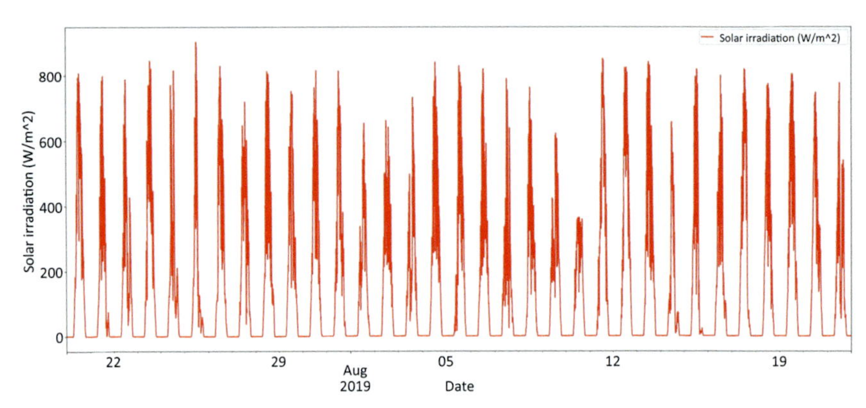

Figure 1.6 The actual data of solar irradiation outside the photovoltaic panel.

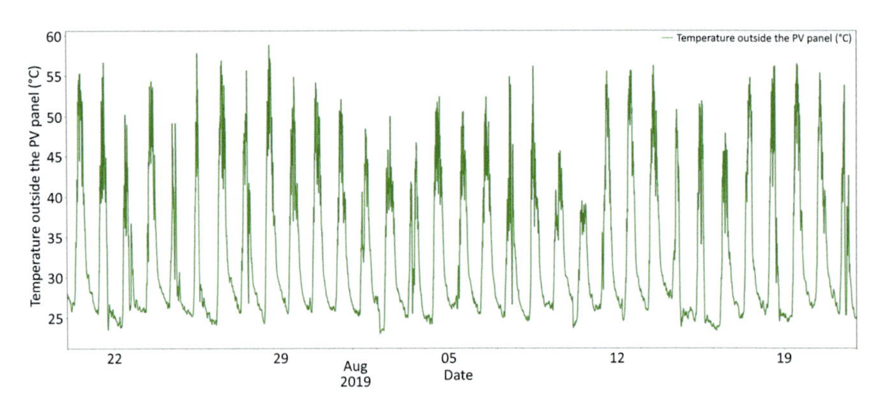

Figure 1.7 The actual data of the temperature outside the photovoltaic panel.

We split the dataset into training and testing subsets as 80%–20%, respectively. retaining 23 days for testing and using 7 days as testing data (validation of model). The full data set of the solar irradiation and temperature is illustrated in Figs. 1.6 and 1.7, respectively. From the data set collected, we can note that the solar irradiation and temperature are in the range of [0, 1000] W/m^2 and [20, 57]°C.

Different ML models were applied and compared to promote this objective. The models considered in this study are RF, LR, KNN, and SVM. The models were executed in Python using a Jupyter Notebook with Scikit-learn library. The error criteria used in this chapter were RMSE.

Tables 1.3 and 1.4 show the comparison of RMSE between the models for forecasting solar irradiation and temperature, respectively. As can

Table 1.3 Root mean square error of solar irradiation for approaches.

	RF	LR	KNN	SVM
RMSE	0.381	0.985	0.933	0.419

Table 1.4 Root mean square error of temperature for approaches.

	RF	LR	KNN	SVM
RMSE	0.454	0.4891	0.4813	0.467

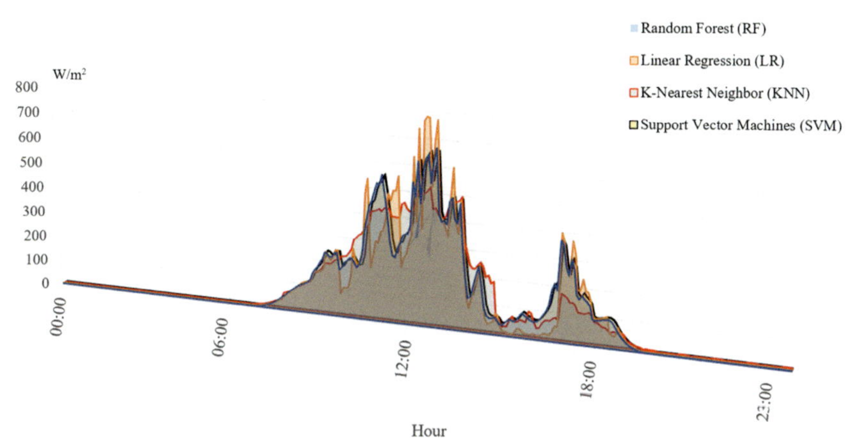

Figure 1.8 Comparison of the proposed models for forecast solar irradiation in daily time.

be seen from these tables, the RF, and SVM less error during testing compared to the LR and KNN, due to their inherent characteristics of carrying initially learned training data. In our review, the RF and SVM show their effectiveness for short-term forecasting of solar irradiation and temperature.

Figs. 1.8 and 1.9 show the comparison of the proposed ML models to forecast the next day of the month of solar irradiation and temperature, respectively.

1.3.1 Recommendation section

1. The performance of the forecasting model is significantly related to the training data set.
2. The RF and SVM methods can promote the forecasting model comparatively high performance with a small data dimension than other

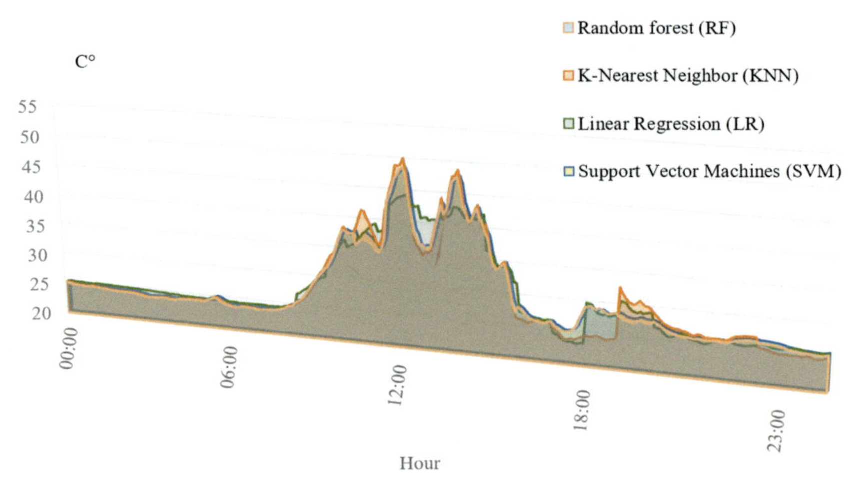

Figure 1.9 Comparison of the proposed models for forecast solar irradiation in daily time.

ML techniques in the proposed study. Moreover, the RF approach has slightly better potential and higher accuracy than the SVM.

3. The RF model and other ML techniques need accuracy, and the optimal parameter is mainly dependent on the size of the trees and leaves. This gives it potential to use the cross-validation for parameter optimization and apply the same optimized parameters to all simulations.

1.4 Conclusion

In the MG, it is important to predict RE power dispatch, and load demand and especially for PV plants. For these objectives, short-term power generation should be predicted for intermittency resources. The PV power dispatch forecasting is correlated to solar irradiation and temperature. In this work, we reviewed some of the ML techniques that have been used to forecast the solar irradiation and temperature.

Our review experiments on real data sets of an installed PV showed that RF and SVM can significantly increase the prediction performance to forecast the solar irradiation and temperature in short terms. It significantly outperforms the results of the typical methods KNN and LR. However, the SVM slightly performed better than RF, due to its inherent characteristics and parameters.

References

[1] M. Shakir, Y. Biletskiy, Forecasting and optimisation for microgrid in home energy management systems, IET Gener. Transm. Distrib. 14 (17) (2020) 3458–3468. Available from: https://doi.org/10.1049/iet-gtd.2019.1285.

[2] Z. Younes, I. Alhamrouni, S. Mekhilef, M. Reyasudin, A memory-based gravitational search algorithm for solving economic dispatch problem in micro-grid, Ain Shams Eng 12 (2021) 1985–1994. Available from: https://doi.org/10.1016/j.asej.2020.10.021.

[3] Younes Zahraoui, M.R. Basir Khan, I AlHamrouni, S. Mekhilef, Current Status, Scenario, and Prospective of Renewable Energy in Algeria: A Review, Energies 14 (2021) 2354. Available from: https://doi.org/10.3390/en14092354.

[4] R. Ahmed, V. Sreeram, Y. Mishra, M.D. Arif, A review and evaluation of the state-of-the-art in PV solar power forecasting : techniques and optimization, Renew. Sustain. Energy Rev. 124 (2020) 109792. Available from: https://doi.org/10.1016/j.rser.2020.109792.

[5] A. Moradzadeh, H. Moayyed, S. Zakeri, B. Mohammadi-Ivatloo, A. Pedro Aguiar, Deep learning-assisted short-term load forecasting forsustainable management of energy in microgrid, Inventions 6 (1) (2021) 15. Available from: https://doi.org/10.3390/inventions6010015.

[6] H. Shuai, H. He, Online scheduling of a residential microgrid via Monte-Carlo tree search and a learned model, IEEE Trans. Smart Grid 12 (2) (2021) 1073–1087. Available from: https://doi.org/10.1109/TSG.2020.3035127.

[7] R. Vincent, M. Ait-Ahmed, A. Houari, M.F. Benkhoris, Residential microgrid energy management considering flexibility services opportunities and forecast uncertainties, Int. J. Electr. Power Energy Syst. 120 (2020) 105981. Available from: https://doi.org/10.1016/j.ijepes.2020.105981.

[8] H. Kanchev, D. Lu, F. Colas, V. Lazarov, B. Francois, Energy management and operational planning of a microgrid with a PV-based active generator for smart grid applications, IEEE Trans. Ind. Electron. 58 (10) (2011) 4583–4592. Available from: https://doi.org/10.1109/TIE.2011.2119451.

[9] H. Sekar, R. Rajashekar, F. Faisal, R. Ganpati, V. Vijayaraghavan, Intelligent dynamic grid forecasting algorithm for a grid-connected solar PV based microgrid, GHTC 2016 – IEEE Global Humanitarian Technology Conference: Technology for the Benefit of Humanity, Conference Proceedings, IEEE, 2016, pp. 421–427. Available from: https://doi.org/10.1109/GHTC.2016.7857315.

[10] E. Scolari, F. Sossan, M. Paolone, Irradiance prediction intervals for PV stochastic generation in microgrid applications, Sol. Energy 139 (2016) 116–129. Available from: https://doi.org/10.1016/j.solener.2016.09.030.

[11] R.H. Inman, H.T.C. Pedro, C.F.M. Coimbra, Solar forecasting methods for renewable energy integration, Prog. Energy Combust. Sci. 39 (6) (2013) 535–576. Available from: https://doi.org/10.1016/j.pecs.2013.06.002.

[12] R. Jane, G. Parker, G. Vaucher, M. Berman, Characterizing meteorological forecast impact on microgrid optimization performance and design, Energies 13 (3) (2020) 577. Available from: https://doi.org/10.3390/en13030577.

[13] Y. Hao, L. Dong, J. Liang, X. Liao, L. Wang, L. Shi, Power forecasting-based coordination dispatch of PV power generation and electric vehicles charging in microgrid, Renew. Energy 155 (2020) 1191–1210. Available from: https://doi.org/10.1016/j.renene.2020.03.169.

[14] R. Sabzehgar, D.Z. Amirhosseini, M. Rasouli, Solar power forecast for a residential smart microgrid based on numerical weather predictions using artificial intelligence methods, J. Build. Eng. 32 (2020) 101629. Available from: https://doi.org/10.1016/j.jobe.2020.101629.

[15] D. Kumar, H.D. Mathur, S. Bhanot, R.C. Bansal, Forecasting of solar and wind power using LSTM RNN for load frequency control in isolated microgrid, Int. J. Model. Simul. 41 (4) (2020) 311–323. Available from: https://doi.org/10.1080/02286203.2020.1767840.

[16] F. Wang, Z. Zhen, B. Wang, Z. Mi, Comparative study on KNN and SVM based weather classification models for day ahead short term solar PV power forecasting, Appl. Sci. 8 (1) (2018) 28. Available from: https://doi.org/10.3390/app8010028.

[17] S. Salcedo-Sanz, C. Casanova-Mateo, A. Pastor-Sánchez, M. Sánchez-Girón, Daily global solar radiation prediction based on a hybrid Coral Reefs Optimization – Extreme Learning Machine approach, Sol. Energy 105 (2014) 91–98. Available from: https://doi.org/10.1016/j.solener.2014.04.009.

[18] J. Shi, W.J. Lee, Y. Liu, Y. Yang, P. Wang, Forecasting power output of photovoltaic systems based on weather classification and support vector machines, IEEE Trans. Ind. Appl. 48 (3) (2012) 1064–1069. Available from: https://doi.org/10.1109/TIA.2012.2190816.

[19] S. Cao, J. Cao, Forecast of solar irradiance using recurrent neural networks combined with wavelet analysis, Appl. Therm. Eng. 25 (2–3) (2005) 161–172. Available from: https://doi.org/10.1016/j.applthermaleng.2004.06.017.

[20] H.T.C. Pedro, C.F.M. Coimbra, Assessment of forecasting techniques for solar power production with no exogenous inputs, Sol. Energy 86 (7) (2012) 2017–2028. Available from: https://doi.org/10.1016/j.solener.2012.04.004.

[21] N. Manoj Kumar, M.S.P. Subathra, Three years ahead solar irradiance forecasting to quantify degradation influenced energy potentials from thin film (a-Si) photovoltaic system, Results Phys. 12 (2019) 701–703. Available from: https://doi.org/10.1016/j.rinp.2018.12.027.

[22] J. Cao, X. Lin, Study of hourly and daily solar irradiation forecast using diagonal recurrent wavelet neural networks, Energy Convers. Manag. 49 (6) (2008) 1396–1406. Available from: https://doi.org/10.1016/j.enconman.2007.12.030.

[23] P. Ramsami, V. Oree, A hybrid method for forecasting the energy output of photovoltaic systems, Energy Convers. Manag. 95 (2015) 406–413. Available from: https://doi.org/10.1016/j.enconman.2015.02.052.

[24] M.R. Zarrindast, M.R. Oveisi, Effects of monoamine receptor antagonists on nicotine-induced hypophagia in the rat, Eur. J. Pharmacol. 321 (2) (1997) 157–162. Available from: https://doi.org/10.1016/S0014-2999(96)00935-1.

[25] M. Aslam, J.M. Lee, H.S. Kim, S.J. Lee, S. Hong, Deep learning models for long-term solar radiation forecasting considering microgrid installation: a comparative study, Energies 13 (1) (2020) 147. Available from: https://doi.org/10.3390/en13010147.

[26] M.C. Kang, J.M. Sohn, J.Y. Park, S.K. Lee, Y.T. Yoon, Development of algorithm for day ahead PV generation forecasting using data mining method, 2011 IEEE 54th International Midwest Symposium on Circuits and Systems (MWSCAS), IEEE, 2011, pp. 1–4. Available from: https://doi.org/10.1109/MWSCAS.2011.6026333.

[27] S. Leva, A. Dolara, F. Grimaccia, M. Mussetta, E. Ogliari, Analysis and validation of 24 hours ahead neural network forecasting of photovoltaic output power, Math. Comput. Simult. 131 (2017) 88–100. Available from: https://doi.org/10.1016/j.matcom.2015.05.010.

[28] A. Dolara, S. Leva, G. Manzolini, Comparison of different physical models for PV power output prediction, Sol. Energy 119 (2015) 83–99. Available from: https://doi.org/10.1016/j.solener.2015.06.017.

[29] K. Mandal, M. Rajkumar, P. Ezhumalai, D. Jayakumar, R. Yuvarani, Improved security using machine learning for IoT intrusion detection system, Mater. Today Proc. (2020). Available from: https://doi.org/10.1016/j.matpr.2020.10.187.

[30] V. Rodriguez-Galiano, M. Sanchez-Castillo, M. Chica-Olmo, M. Chica-Rivas, Machine learning predictive models for mineral prospectivity: an evaluation of neural networks, random forest, regression trees and support vector machines, Ore Geol. Rev. 71 (2015) 804–818. Available from: https://doi.org/10.1016/j.oregeorev.2015.01.001.
[31] W. Zhang, C. Wu, Y. Li, L. Wang, P. Samui, Assessment of pile drivability using random forest regression and multivariate adaptive regression splines, Georisk 15 (1) (2021) 27–40. Available from: https://doi.org/10.1080/17499518.2019.1674340.
[32] M.B. Abidine, B. Fergani, Human activity recognition in smart home using prior knowledge based KNN-WSVM model, in: M. Hatti (Ed.), Artificial Intelligence in Renewable Energetic Systems. ICAIRES 2017. Lecture Notes in Networks and Systems, vol. 35, Springer, Cham, 2018, pp. 15–24. Available from: https://doi.org/10.1007/978-3-319-73192-6_2.
[33] M.Q. Raza, M. Nadarajah, C. Ekanayake, On recent advances in PV output power forecast, Sol. Energy 136 (2016) 125–144. Available from: https://doi.org/10.1016/j.solener.2016.06.073.

CHAPTER TWO

Generators' revenue augmentation in highly penetrated renewable M2M coordinated power systems

Deepak Yadav[1], Saad Mekhilef[1,2,3], Brijesh Singh[4], Muhyaddin Rawa[3], Yusuf Alturki[5] and Abdullah Abusorrah[3]

[1]Power Electronics and Renewable Energy Research Laboratory, Department of Electrical Engineering, University of Malaya, Kuala Lumpur, Malaysia
[2]School of Science, Computing and Engineering Technologies, Swinburne University of Technology, Hawthorn, VIC, Australia
[3]Center of Research Excellence in Renewable Energy and Power Systems, King Abdulaziz University, Jeddah, Saudi Arabia
[4]Department of Electrical and Electronics Engineering, KIET Group of Institutions, Ghaziabad, India
[5]Electrical and Computer Engineering Department, King Abdulaziz University, Jeddah, Saudi Arabia

2.1 Introduction

The rapid growth in demand and increasing global warming inducing higher investment in clean and affordable energy resources. Along with easy installations, the renewable resources of solar Photovoltaic (PV) have been encouraged. The infrastructural and policies of renewable energy resources are highly resilient to promote PV power generations. The affordability of solar PV driving the major commencements of the projects around the globe. Investors are very keen to grab profit with the future energy market [1]. Although the higher PV power injections in the power system arise some major challenges. The PV units increase the voltage levels locally beyond the acceptable limits [2–4]. The higher penetration of PV units (large scale) may cause excessive power generation during peak hours, and the demands are also less during the mid-day (peak hour of PV units). This overgenerated power without any intervention may damage the electrical equipment and rotating machines connected to the network. The load bus of the network can be penetrated up to 120%–250% maximum power point of the minimum daytime

Applications of AI and IOT in Renewable Energy.
DOI: https://doi.org/10.1016/B978-0-323-91699-8.00002-4

demand of the corresponding bus. This level of PV penetration is desirable with the assistance of advanced converters without overvoltage issues [5–7]. Furthermore, the power systems market exhibits some other challenges that is power system operation, reliability, the optimal transmission of power, security, economic dispatch, etc. Consideration of optimum management strategies of highly injected PV systems is one of the most complex and challenging issues in the modern power system. The major challenges that arise due to higher penetration of PV systems in the network can be prevented by an efficient market-to-market (M2M) interconnection.

In the article [8], it has been explained that the PV systems are generating extremely low price electricity. Therefore, it has become complex decision-making in the long-term to operate the conventional generators with a profitable business. Further, the overgeneration would lead to imposing the overflow (in other words- congestion in transmission lines) and leads to a rise in the voltages of various nodes into the network [9,10]. The optimal power flow (OPF) based operational strategy may provide a better solution to operating the power systems market with higher efficiency. Independent system operators (ISOs) and transmission system operators have already proposed guidelines for the interregional/ M2M electricity exchanges. Most of them have not clarified the nodal properties being involved in the power trade. United State's Federal Energy Regulatory Commission (FERC) has instituted guidelines in order no. 1000, although the interconnections of ISOs/RTOs are based on financial obligations. Further PJM operators have proposed that the system should be tested for the commenced projects to analyze the technical factors of the systems [11,12].

In this work, the nodal properties' effect on the systems price and revenue have been analyzed. Market operators and researchers are striving to develop advanced policies to upturn the earnings. The authors in the article [13] have discussed the impacts of carbon taxes and long-term policies to boost renewable penetration through various investments or production credits. Further, the article has analyzed the different policies that is solar investment tax credit (ITC), the production tax credit (PTC), feed-in-tariffs, renewable portfolio standard (RPS), and carbon tax/cap-and-trade program. In Xuet al. [14] conditional value-at-risk method incorporating scenario analysis method (to deal with uncertainty parameters) has been used to analyze the potential risk to purchase the electricity in the spot markets. The results of the proposed method have shown the revenue increment in the spot market.

In this work, the locational marginal prices (LMPs) of generators' nodes have been considered as the offer price. Therefore, the revenue to the generator would be the unit-commitment multiple of the offer price. The generators' unit-commitments have been allotted using SC-OPF. Yadav et al. [15] have studied the interconnections method based on nodal properties during the overgeneration scenario. And it is observed that the contiguous interconnection of PV systems to loads would provide an economic dispatch of electricity. The Interior-Point (IP) optimization technique based on Karush–Kuhn–Tucker (KKT) optimality conditions has been used to obtain the optimal solutions. The optimal schedules of the generators have been obtained using IP-SCOPF.

2.2 Problem formulation

The central objectives of this chapter are to operate the highly renewable resources penetrated network to obtain economic dispatch and augment the generators' revenue in the networks. The solutions to the formulated problem have been obtained from IP-OPF.

The objective function along with associated security constraints are represented as:

$$\min f\left(P_{g_i}\right) = \sum_{u=1}^{U}\sum_{i=1}^{G}\left\{C_i\left(P_{g_i}\right)\right\} \tag{2.1}$$

s.t.

$$\sum_{u=1}^{U}\left\{\sum_{i=1}^{G}P_{g_i}\right\} - \sum_{i=1}^{D}P_{d_i} - P_{\text{loss}} = 0 \tag{2.2}$$

$$\underline{P_{g_i}} \le P_{g_i} \le \overline{P_{g_i}};\text{g}_i \in \text{Coventional Gen} \tag{2.3}$$

$$\underline{Q_{g_i}} \le Q_{g_i} \le \overline{Q_{g_i}};\text{g}_i \in \text{Coventional Gen} \tag{2.4}$$

$$P_{g_i} = P_{g_i}{}^{\text{MPP}};\text{g}_i \in \text{PV units} \tag{2.5}$$

The negligible value of reactive power output of PV units (∼20% of active power output) is not considerable for this study.

$$\underline{V_i} \leq V_i \leq \overline{V_i} \tag{2.6}$$

$$S_{\text{ik}} \leq \overline{S_{\text{ik}}} \tag{2.7}$$

The active power LMPs have been considered in this work, the LMPs for reactive power have been omitted due to negligible values. The LMP values have been obtained from the Lagrange active power balancing equations' and it is the Lagrange multiplier at each bus.

2.2.1 Locational marginal prices expression

The partial derivative of the with respect to the state variables (angles and voltages) of the Lagrange function must be 0 which is the imperative power flow optimality condition. The aforesaid condition yields the LMPs equation for the bus i_s.

$$\lambda_{P_i} = \frac{\partial C_s(P_s)}{\partial P_s} - \frac{\partial P_{\text{loss}}}{\partial P_i}\frac{\partial C_s(P_s)}{\partial P_s} - \frac{\partial \mathcal{H}^T}{\partial P_i}\mu \tag{2.8}$$

Eq. (2.8) comprises three factors, energy, loss, and congestion, respectively.

2.3 Algorithm

$$\max(\lambda_{P_i} * \mathsf{P}_{\mathsf{g_i}}); \lambda_{\mathsf{P_i}} \in \mathsf{G} \tag{2.9}$$

s.t.

constraint (2)−(6)

Condition 1 − subject to condition:

$$\sum_{u=1}^{U}\left\{\sum_{i=1}^{G}(P_{g_i}) - \sum_{i=1}^{D}P_{d_i} - P_{\text{loss}}\right\} = 0, \text{ M2M tie-line status: off}$$

$$S_{ik} \leq \overline{S_{ik}} \text{and} S_{T_l} = 0 \tag{2.10}$$

Condition 2 − subject to overgeneration condition:

$$\sum_{u=1}^{U}\left\{\sum_{i=1}^{G}(P_{g_i}) - \sum_{i=1}^{D}P_{d_i} - P_{loss}\right\} > 0, \text{ M2M tie-line status: on}$$

$$S_{ik} \leq \overline{S_{ik}}; i \in g^{PV}; k \in d \text{and} S_{T_l} \leq \overline{S_{T_l}} \tag{2.11}$$

Eq. (2.11) exhibits the contiguous interconnection of identified PV systems to the identified demand nodes. Similarly, the other two cases have been analyzed in this work.

2.4 Interior-point technique and KKT condition

The current problem of optimizing is nonlinear. Therefore a conventional nonlinear optimization method was applied for solving the nonlinear problem. One of the powerful optimization algorithms for solving nonlinear optimization problems is Karmarkar's IP method. Conventional optimization techniques, overall, involved several iterations and took longer to find an optimal solution. Nevertheless, this is a recursive algorithm that cuts through the interior of the optimal solution in IP-based optimization. Consequently, with a significant iteration, the IP method brings a fast result. The IP approach is therefore more suited designed for excessive sizable LPs [16]. The core conception of the IP and its approach is the following:

Let's say, the problem is outlined as

$$\begin{gathered} \min e(y) \\ \text{s.t.} f(y) = 0 \\ \text{gl} \leq g(y) \leq \text{gu} \end{gathered} \tag{2.12}$$

where $e(y)$ is an objective function. $f(y)$ is a set of nonlinear equality constraints (Power balance) and $\text{gl} \leq g(x) \leq \text{gu}$ is a set of nonlinear inequality constraints; Lagrange function is developed for (2.12) redrafted as for OPF solutions applying the IP technique.

$$L_g = e(y) - x^T f(y) - z^T [g(y) - l - gl] - w^T [g(y) + u - gu] - \gamma \sum_{i=1}^{r} \ln l_i - \gamma \sum_{i=1}^{r} \ln u_i \tag{2.13}$$

where -x, z, and w: Lagrange multipliers for equality and inequality constraints correspondingly; l_i and u_i: slack variables; γ: boundary restriction. Now, afterward authentication of the KKT condition for (2.13), the Jacobean matrices J_e, J_f, and J_g and Hessian matrices H_e, H_f, and H_g are obtained, respectively, from $e(y)$, $f(y)$ and $g(y)$. Using a reduced Newton form, a decomposed linear equation is subsequently obtained for (2.13).

2.4.1 Karush–Kuhn–Tucker conditions

Consider a general optimization problem for a first- or second-order situation,

$$\min e(y)$$

subject to

$$f_i(y) \leq 0, \ i = 1, 2, 3, \ldots, p$$

$$g_i(y) = 0, \ i = 1, 2, 3, \ldots, q. \tag{2.14}$$

Assuming that y^* is a local minimum for the certain case and that is a stable constraints, then there occur exclusive vectors μ_i^* ($i = 1, 2, 3, \ldots, p$) and λ_i^* ($i = 1, 2, 3, \ldots, q$), known as KKT multipliers [17], are present, so that

$$\nabla e(y^*) + \nabla f_i(y^*)^T \mu^* + \nabla g_i(y^*)^T \lambda^* = 0, \tag{2.15}$$

$$\mu^* \geq 0, \tag{2.16}$$

$$f_i(y*) \leq 0, \tag{2.17}$$

$$g_i(y*) \geq 0, \tag{2.18}$$

$$(\mu*)^T f_i(y*) = 0. \tag{2.19}$$

The conditions (2.15)–(2.19) are referred to as the conditions of KKT. To achieve the optimal solutions for pseudo optimization, all such conditions must be satisfied. The KKT conditions are adequate for global optimality for the convex optimization problems and affinity constraints.

2.4.2 Solution algorithm

The key steps for the OPF based IP solution are as follows.

Step 1) Initialization: provide the formulated objective function with initial values.

Step 2) Formulate Jacobean and Hessian matrix of the objective function, equality, and inequality constraints: J_e, J_f, J_g, H_e, H_f, and H_g.

Step 3) To solve this linear system, formulate the linear equation for (2.12), applying a predictor-corrector technique. If the conditions for convergence are met, then stop; otherwise, go to step 2.

2.5 Test results and discussion

This work has been examined for M2M analysis using standard IEEE-118 and IEEE-57 bus systems [18,19]. The maximum power will be assigned to PV units during the rescheduling of the generators due to low-cost generations. Since the analysis was performed for the mid-day overgeneration scenario due to low demand and peak PV generation.

The PV penetrations have been taken 120% MPP of minimum day time load (MDTL). The case studies have been analyzed for a single period.

The maximum output for the respective penetration percentage (PoP) is expressed as a function:

$$MPPP_i = PoP * MDTL \text{ at load bus } i \tag{2.20}$$

Where the active power output of PV units at bus i is P_i and PoP is 120% of MDTL.

Case 1: Slack to Slack bus interconnection

Since the slack buses provide higher flexibilities to the system, therefore the revenue augmentation analysis using this interconnection would provide better policy-making decisions.

Case 2: Identified PV units to identified load buses

This M2M interconnection will enable a direct path to transmit the power from PV units to the load. Hence, the analysis for this interconnection may provide a better solution for future grids.

Case 3: Fictitious bus to slack bus

In general, fictitious buses are used to accumulate the generated power or load aggregation in the system. In this work, the power of the specified PV units will thus be accumulated in a fictitious bus that would have been interconnected to other network's slack buses.

Case 4: Fictitious bus to fictitious bus

In this situation, the outputs of the PV units are accumulated via a fictitious bus and then transferred to another network's fictitious bus. The fictitious bus of receiving network aggregates loads of the specified load buses.

The load profile for the standard IEEE-118 and IEEE-57 bus systems have been shown in Figs. 2.1 and 2.2.

Fig. 2.1 illustrates that the overall demand for the 20 MW lower limit of conventional generators has been surpassed by cumulative generations.

Therefore to manage the excess power and generate revenue for the generators the M2M interconnections have been tested using standard IEEE-57 bus system with the standard IEEE-118 bus system. The reactive power profile has also been considered in the test to assist the voltage profile in the system. Although, the analysis has been performed for the active power profiles.

Fig. 2.3 shows the active power generation profiles for the different types of M2M interconnections. The generation profile shows that the IEEE-118 bus network's conventional generator outputs are higher for the direct PV units interconnections with defined load buses, while the IEEE-57 bus network's conventional generator outputs are minimum.

From Fig. 2.4, it can be observed that the direct PV units interconnections with defined load buses yield the lowest costs for the power generations.

Since the LMPs at the generating buses have been considered as the offer costs. Therefore Figs. 2.5 and 2.6 shows that the direct PV units interconnections with defined load buses have a slightly higher value of LMP and higher revenue generated in the dispatching network (IEEE-118 bus systems). Further, PV units' revenue is also higher in the case of direct interconnections of PV units to loads.

Correlated to other policies to increase the revenue of generators (conventional and renewable resources), this work has analyzed the

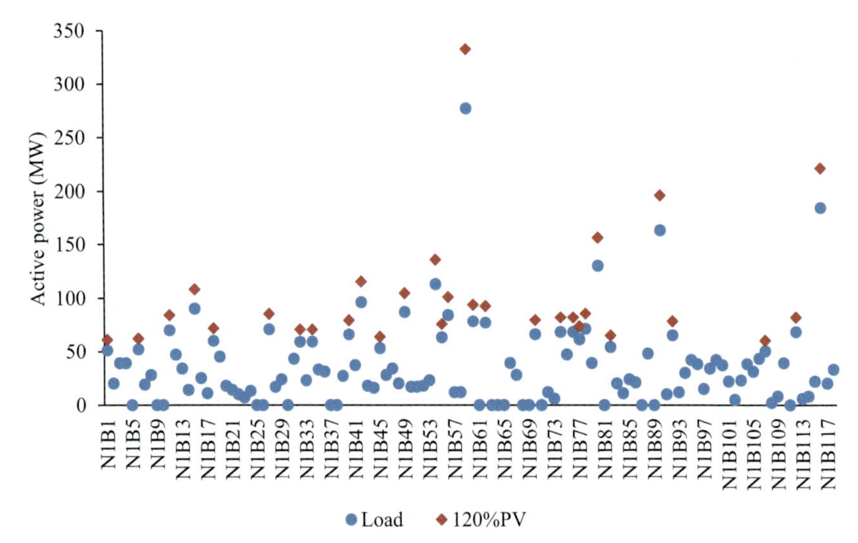

Figure 2.1 Active power demand and photovoltaic outputs in IEEE-118 bus system.

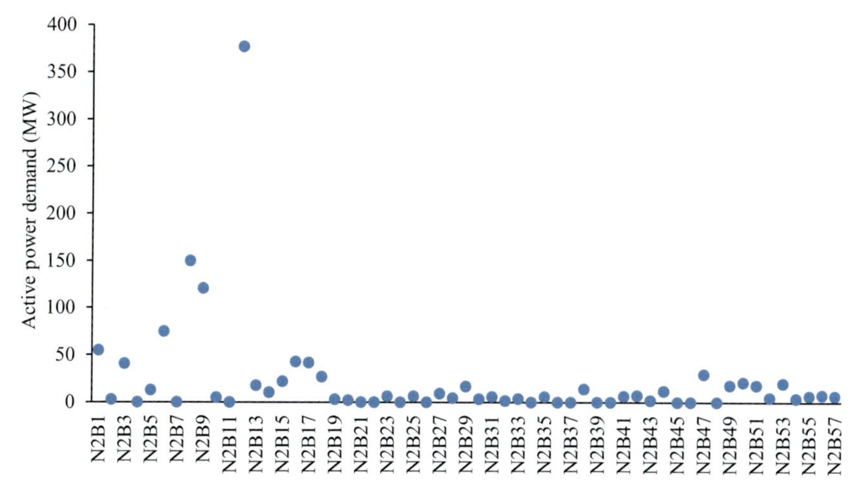

Figure 2.2 Active power demand in IEEE-57 bus system.

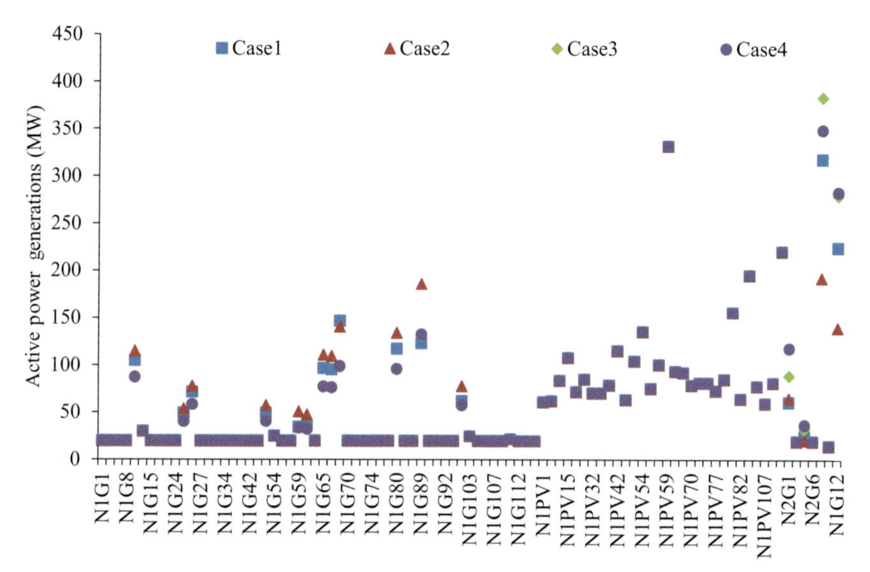

Figure 2.3 Active power (MW) generation profiles.

interconnection methods as per FERC/PJM road-map for novel overgeneration scenarios considering network expansions. This work has calculated the revenue generation in the spot market and has not dealt with carbon tax/cap-and-trade programs, RPS, ITC, PTC, and feed-in-tariffs.

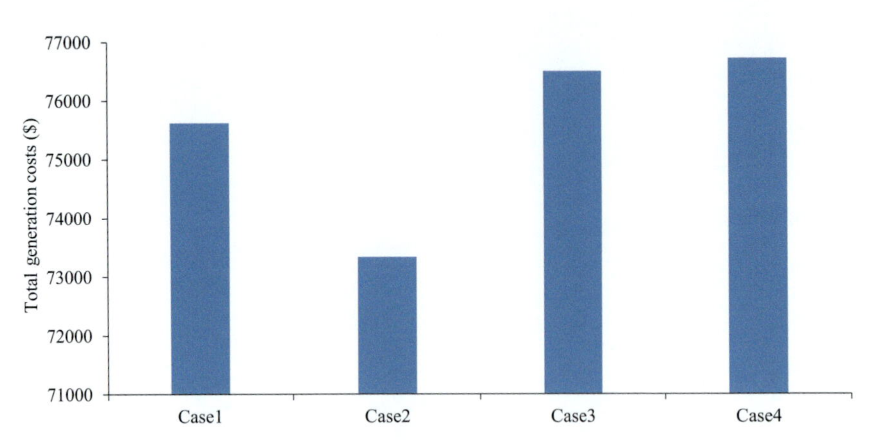

Figure 2.4 Total generation costs for all cases ($).

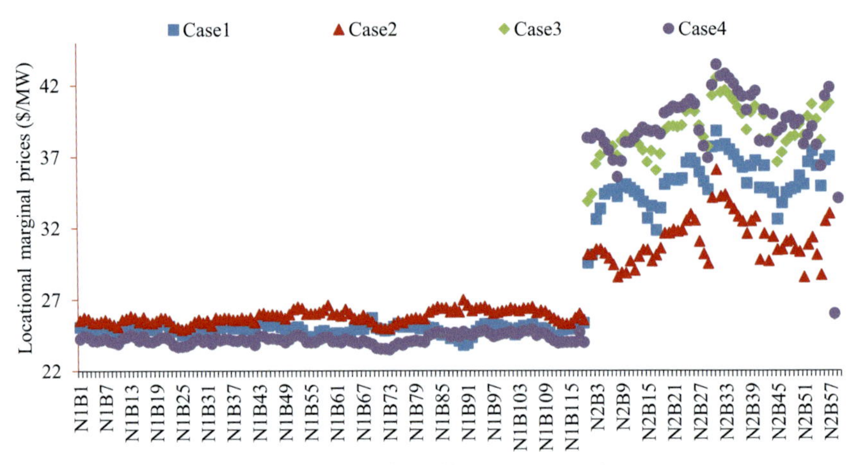

Figure 2.5 Locational marginal prices for different case studies.

The proposed market model and interconnecting conditions are shown by a graphical depiction in Fig. 2.7. U1 and U2 are two networks in the system, and the U1 is highly penetrated with PV units. In the course of off-peak PV availability, these two networks attain the market equilibrium within individual markets. While the markets coordinate with each other over the excess power outputs.

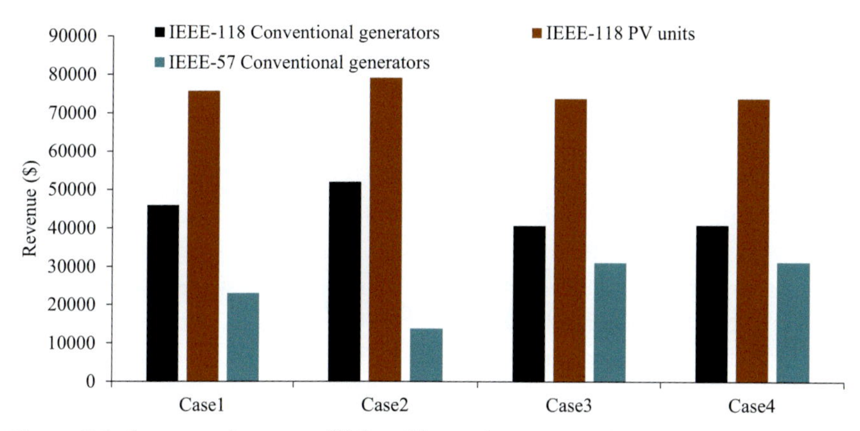

Figure 2.6 Generators' revenue ($) for different interconnections.

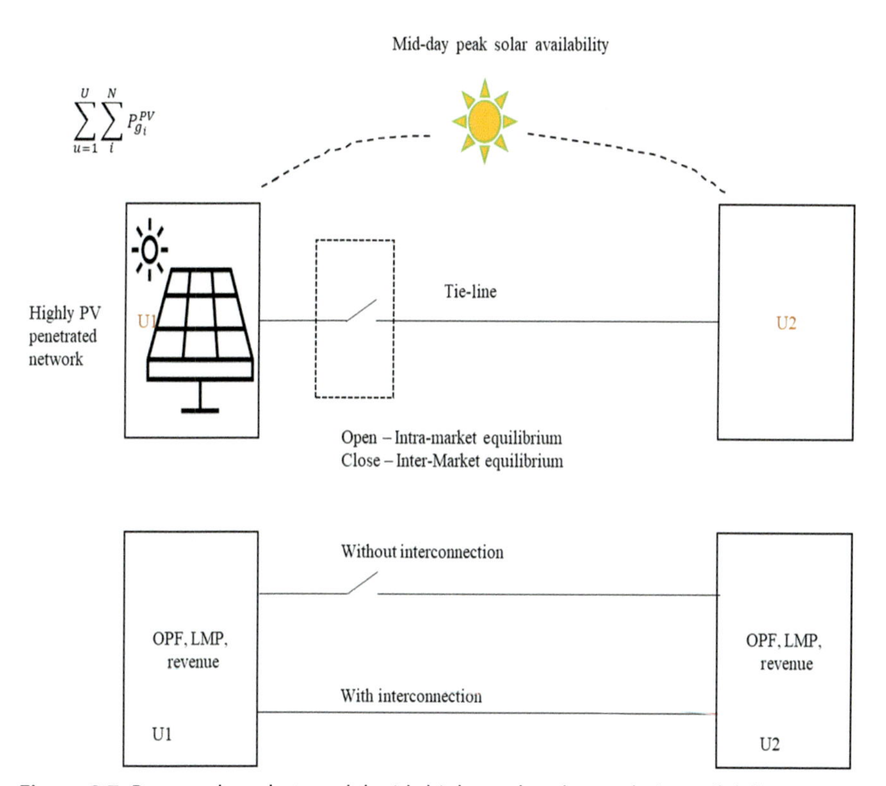

Figure 2.7 Proposed market model with higher solar photovoltaic availability.

2.6 Conclusion

This work has analyzed the different interconnections in an M2M of the power system. The cases have been analyzed for the excess power generations a network and to manage this excess generated power it is interconnected to another network. Therefore to augment the revenue of generators with higher PV penetration this work has studied all interconnections. Further, the LMP values at respective generator buses have been considered as the offer prices. The IP-OPF solutions have provided that the direct PV units interconnections with defined load buses generate more revenue for the conventional and PV generators in the dispatching network (market). The LMP values and conventional generators outputs are higher though. Although the total generation costs are much lower, therefore with an effective price/cost distribution between the different markets these interconnections may provide a long-term electricity exchange with higher effectiveness. The price/cost distribution mechanism between markets would be studied in future work.

References

[1] R. Fu, D. Feldman, R. Margolis, United States Solar Photovoltaic System Cost Benchmark: Q1 2018, NREL, Golden, CO, 2018.

[2] M. Obi, R. Bass, Trends and challenges of grid-connected photovoltaic systems - a review, Renew. Sustain. Energy Rev. 58 (2016) 1082–1094. Available from: https://doi.org/10.1016/j.rser.2015.12.289.

[3] N.M. Haegel, et al., Terawatt-scale photovoltaics: transform global energy, Science 364 (6443) (2019) 836–838. Available from: https://doi.org/10.1126/science.aaw1845.

[4] K.N. Nwaigwe, P. Mutabilwa, E. Dintwa, An overview of solar power (PV systems) integration into electricity grids, Mater. Sci. Energy Technol. 2 (3) (2019) 629–633. Available from: https://doi.org/10.1016/j.mset.2019.07.002.

[5] A. Hoke et al., Inverter Ground Fault Overvoltage Testing. National Renewable Energy Lab (NREL), Golden, CO. https://doi.org/10.2172/1215244.

[6] R. Mcallister, D. Manning, L. Bird, M. Coddington, C. Volpi, New Approaches to Distributed PV Interconnection: Implementation Considerations for Addressing Emerging Issues, National Renewable Energy Lab (NREL), Golden, CO. https://doi.org/10.2172/1505553.

[7] B. Palmintier, et al., Final Technical Report: Integrated Distribution-Transmission Analysis for Very High Penetration Solar PV, National Renewable Energy Lab (NREL), Golden, CO, 2016.

[8] A. Rai, O. Nunn, On the impact of increasing penetration of variable renewables on electricity spot price extremes in Australia, Econ. Anal. Policy 67 (2020) 67–86. Available from: https://doi.org/10.1016/j.eap.2020.06.001.

[9] P. Denholm, K. Clark, M. O'Connell, On the Path to SunShot - Emerging Issues and Challenges in Integrating High Levels of Solar into the Electrical Generation and Transmission System, 2016. doi:10.2172/1344204.
[10] M.A. Khan, N. Arbab, Z. Huma, Voltage profile and stability analysis for high penetration solar photovoltaics, Int. J. Eng. Work 5 (5) (2018) 109–114.
[11] PJM Interconnection, L.L.C., 155, vol. 250, no. 2014, 2016.
[12] Amended and Restated Operating Agreement of PJM Interconnection, L.L.C., no. 24, 2011.
[13] T. Levin, J. Kwon, A. Botterud, The long-term impacts of carbon and variable renewable energy policies on electricity markets, Energy Policy 131 (2019) 53–71. Available from: https://doi.org/10.1016/j.enpol.2019.02.070.
[14] W. Xu, P. Zhang, D. Wen, Decision-making model of electricity purchasing for electricity retailers based on conditional value-atRisk in day-ahead market, in: 2020 12th IEEE PES Asia-Pacific Power and Energy Engineering Conference (APPEEC), 2020, pp. 1-5, doi: 10.1109/APPEEC48164.2020.9220431.
[15] D. Yadav, S. Mekhilef, B. Singh, M. Rawa, Analysis of market to market interconnection points during overgeneration scenario in a market, in: 2020 IEEE 5th International Conference on Computing Communication and Automation (ICCCA), 2020, pp. 774–779, doi: 10.1109/ICCCA49541.2020.9250727.
[16] D. Yadav, A.S. Chauhan, B. Singh, Contingency analysis and security constraint based optimal power flow in power network, in: 2018 3rd International Innovative Applications of Computational Intelligence on Power, Energy and Controls with their Impact on Humanity (CIPECH), 2018, pp. 210–214, doi: 10.1109/CIPECH.2018.8724298.
[17] Y.C. Wu, A.S. Debs, R.E. Marsten, A direct nonlinear predictor-corrector primal-dual interior point algorithm for optimal power flows, IEEE Trans. Power Syst. 9 (2) (1994) 876–883. Available from: https://doi.org/10.1109/59.317660.
[18] R. Christie, Power systems test case archive, 118 Bus Power Flow Test Case, University of Washington, Seattle, WA, 1993. [Online]. Available from: <https://labs.ece.uw.edu/pstca/pf118/pg_tca118bus.htm> (accessed 17.01.21).
[19] R. Christie, Power Systems Test Case Archive, 57 Bus Power Flow Test Case, University Washington, Seattle, WA, 1993. [Online]. Available from: <http://labs.ece.uw.edu/pstca/pf57/pg_tca57bus.htm> (accessed 18.01.21).

CHAPTER THREE

Intelligent supervisory energy-based speed control for grid-connected tidal renewable energy system for efficiency maximization

Youcef Belkhier[1], Mohamed Dahman Alshehri[2] and Rabindra Nath Shaw[3]

[1]Department of Computer Science, Faculty of Electrical Engineering, Czech Technical University in Prague, Prague, The Czech Republic
[2]Department of Computer Science, College of Computers and Information, Taif University, Taif, Saudi Arabia
[3]Department of International Relations, Bharath Institute of Higher Education and Research (Deemed to be University), Chennai, India

3.1 Introduction

The tidal energy that results from the transformation of the kinetic energy of marine currents into electrical energy through tidal turbines, has gained increasing attention in recent years due to the advantages of being a clean source of renewable energy and highly predictable, compared to its predecessor [1,2]. It is therefore a marine transposition of the wind rotor that recovers the kinetic energy of the wind. The parallel that can be established between the two technologies can be found in the first place in the similar designs that have been adopted. The use of permanent magnet synchronous generator (PMSG) in tidal turbine system has high potential due to its reliability, increased energy, reduced failure and possibility to eliminate the gearbox which lead to low maintenance and enable to the PMSG to be very favorable in marine current applications [2]. However, the controller computation for the PMSG is still a challenging work, due to unknown modeling error, external disturbances and parameter uncertainties. Furthermore, DC-link overvoltage control, reactive power support, efficiency of the power converter, and fault ride-through capability are the important requirement in the grid-connected tidal energy systems for a

Applications of AI and IOT in Renewable Energy.
DOI: https://doi.org/10.1016/B978-0-323-91699-8.00003-6

reliable and efficient electrical energy [3]. Several research works dedicated to the nonlinear control of PMSG have appeared in the literature and in the industry. In Gu et al. [4], a sliding mode control (SMC) combined with a fuzzy method is investigated. However, the robustness of the proposed strategy under time-varying parameters have not been investigated. In Othman [5], a Jaya-based sliding mode method is proposed. However, the combined strategy increases the costs and maintenances time of the closed-loop. A hybrid controller was developed in Yin and Zhao [6] for the maximum power extraction of new hydrostatic tidal turbine. SMC method in Zhou et al. [7]. However, the authors have not investigated sudden change in the behavior of the tidal current and time-varying parameters. The same system was adopted in Zhou et al. [8], where the SMC control has been replaced by the proposed novel active disturbance rejection control (ADRC) method. The ADRC shows a clear improvement in the control performance compared to that of the SMC and the classical proportional-integral (PI) controls. A novel Q-network algorithm combined with Tilt-based fuzzy cascaded strategy is proposed in Sahu et al. [9]. An SMC with magnetic equivalent circuit is developed in Toumi et al. [10], here also however, the authors have not investigated the robustness of the proposed strategy under time-varying parameters. A linear quadratic control in Gaamouche et al. [11]. Perturb and observe algorithm in Moon et al. [12]. However, the majority of these controls are signal based and do not takes into account the physical structure of the synchronous machine during the controller design, as mentioned in Yang et al. [13].

Within, in this work, a novel controller method, based on the passivity concept that track velocity, and maintain this one operating at the optimal torque, is proposed. Inherent advantages of the passivity-based control (PBC) method also called "Energy-based control" are that the nonlinear properties are not canceled but compensated in a damped way [14,15]. Several researches work about the passivity control dedicated to the PMSG have appeared in the literature and in the industry. A PBC associated with a sliding mode controller is adopted in Yang et al. [16]. However, the proposed strategy uses fixed gains which are difficult to compute when the system is under parameter uncertainties. In Subramaniam and Joo [17] a hybrid PBC, SMC and, fuzzy control is developed. However, the control design is very complex to implement. Passivity-based linear feedback control is explored in Yang et al. [18]. However, nonlinear properties and the robustness due to parameter changes of the PMSG have not been evaluated.

The present paper investigates a novel passivity-based on a combined speed control and fuzzy logic approach to design an optimal controller for the PMSG. The main objectives of this study consist into two main parts: extract the maximum marine current power through controlling the rotational of the PMSG and transfer this power to the grid-side converter (GSC), for this a new controller is proposed. The task of the second one is to regulate the reactive power and DC-link voltage at their prefault values to guarantee efficient, secure, and reliable electricity, indeed any possible disturbances related to the machine-side converter (MSC) such as sudden changes in tidal velocity, parametric uncertainties and PMSG nonlinear properties. Thus a classical PI controller is adopted by the GSC, that is, to enhance the overall performances of the closed-loop. A special focus is given to the MSC, as it is the bridge between the tidal turbine and the grid, that is, by taking into account the entire dynamic of the PMSG when synthesizing the new proposed control strategy. Furthermore, the robustness against parameter changes has been taken a special attention.

The main originality and contribution of the present work over the related papers in the literature are clearly summarized as given below:

1. A novel PBSC combined with fuzzy logic control for optimal performance of a PMSG is proposed and for the improvement of the power quality transferred to the grid. The PBSC is adopted to design the controller law, in order to guarantee a fast convergence of the closed-loop system.
2. The fuzzy logic controller is selected which makes the proposed strategy intelligent to compute the damping gains to make the closed-loop passive and approximate the unstructured dynamics of the PMSG and thus robustness property of the closed-loop system is considerably increased.
3. The essential characteristic of this approach is the extremely reduced number of the fixed gains used by the proposed strategy which avoid its sensitivity to parameter uncertainties, which highly improve the robustness property and global stability of the system.
4. Extensive numerical investigations are made to demonstrate the robustness the proposed approach against parameter changes and external disturbances. Also, extensive comparison with other nonlinear controls is provided to highlight the superiority of the proposed Fuzzy supervisory-PBSC (FS-PBSC).

The present paper is organized by the present form: in Section 3.2, the system description is established. Sections 3.3 and 3.4 deals with the proposed strategy design procedure. Concerning Section 3.5, the controller strategy applied to GSC is formulated. In Section 3.6, extensive numerical investigation and experimental validation of the proposed candidate strategy. Finally, Section 3.7 deal with main conclusions of the present paper.

3.2 Marine current conversion system modeling

The configuration of the investigated conversion system with Matlab/Simulink is presented in Fig. 3.1, is composed by a tidal turbine, PMSG, AC-DC-AC converter, and the grid. The proposed strategy is applied to the MSC to regulates the produced power via the generator which, while the network receives only the active power using the classical PI method represents the aim of the GSC.

3.2.1 Tidal turbine model

The mathematical model of the tidal energy that can be transformed by the turbine is given by Othman [5] and Khefifi [19]:

$$P_m = \frac{1}{2}\rho C_p(\beta, \lambda) A v_s^3 \tag{3.1}$$

$$T_m = \frac{P_m}{\omega_m} \tag{3.2}$$

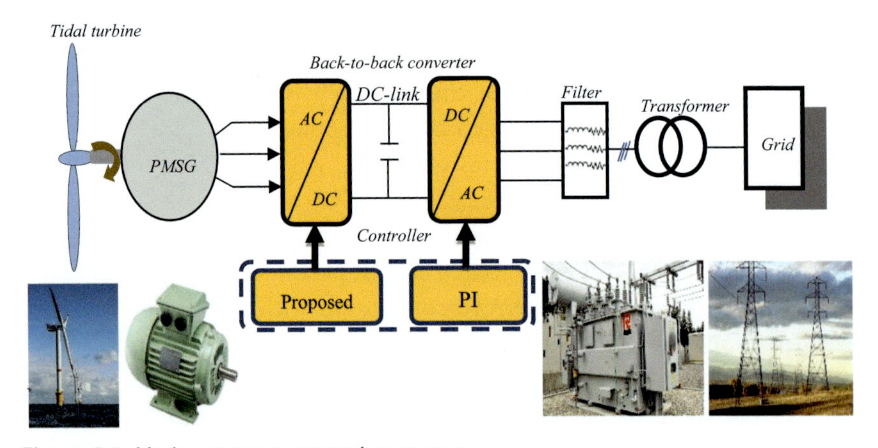

Figure 3.1 Marine current conversion system.

$$C_p(\beta,\lambda) = 0.5\left(\frac{116}{\lambda_i} - 0.4\beta - 5\right)e^{-\left(\frac{21}{\lambda_i}\right)} \tag{3.3}$$

$$\lambda_i^{-1} = (\lambda + 0.08\beta)^{-1} - 0.035\left(1+\beta^3\right)^{-1} \tag{3.4}$$

$$\lambda = \frac{\omega_t R}{v_s} \tag{3.5}$$

where, v_s denotes the marine current speed, β denotes the pitch angle, ω_t denotes the turbine speed, R denotes the radius of the blades, C_p represents the power coefficient, λ denotes tip-speed ratio, ρ represents marine current density, and A represents the swept area of the blades.

3.2.2 Permanent magnet synchronous generator modeling

To design the proposed strategy, the PMSG model in dq-fram is concedired, expressed as [1,20]:

$$v_{dq} = R_{dq}\, i_{dq} + L_{dq}\frac{di_{dq}}{dt} + p\omega_m \Im\left(L_{dq}\, i_{dq} + \psi_f\right) \tag{3.6}$$

$$J\frac{d\omega_m}{dt} = T_m - T_{em} - f_{fv}\omega_m \tag{3.7}$$

$$T_{em} = \frac{3}{2}p\psi_{dq}\Im i_{dq} \tag{3.8}$$

Where, $i_{dq} = \begin{bmatrix} i_d \\ i_q \end{bmatrix}$ represents the stator current vector, T_{em} denotes the electromagnetic torque, $L_{dq} = \begin{bmatrix} L_d & 0 \\ 0 & L_q \end{bmatrix}$ represents the matrix of the stator induction, f_{fv} represents the viscous coefficient, θ_e represents the electrical angular, $v_{dq} = \begin{bmatrix} v_d \\ v_q \end{bmatrix}$ denotes the voltage stator vector, $R_{dq} = \begin{bmatrix} R_S & 0 \\ 0 & R_s \end{bmatrix}$ represents the stator resistance matrix, $\psi_{dq} = \begin{bmatrix} \psi_d \\ \psi_q \end{bmatrix} = \begin{bmatrix} L_d i_d + \psi_f \\ L_q i_q \end{bmatrix}$ is the flux linkages vector, and ω_m denotes the PMSG speed.

3.3 Control of the permanent magnet synchronous generator using passivity method

To design of the proposed FS-PBSC several steps need to be verified: Firstly, it is required to demonstrate the passivity property of the PMSG model such that the proposed method can be applied. Secondly, the PMSG need to decomposed into two passive subsystems with negative feedback. Finally, the nondissipative terms in the PMSG model need to be identified to compute a controller with simple structure. The controller design process is depicted in Fig. 3.2. Two main parts can be distinguished in the investigated candidate process: the first step consists in design of the reference current through and then the controller law computed using the developed FS-PBSC strategy.

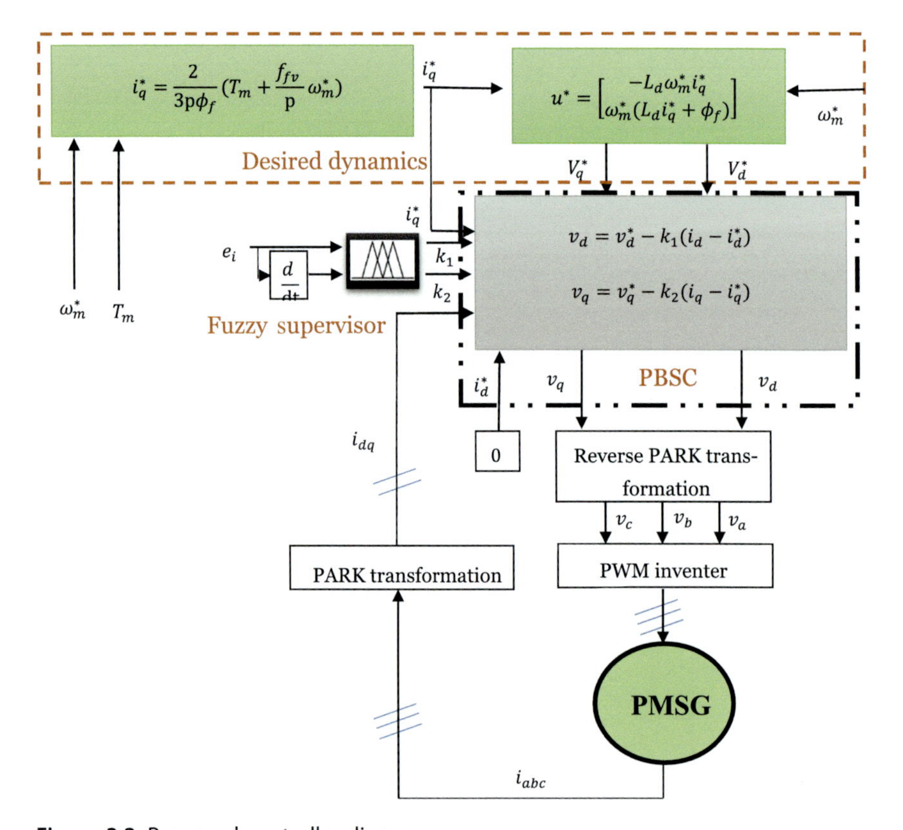

Figure 3.2 Proposed controller diagram.

3.3.1 Permanent magnet synchronous generator dq-model interconnected subsystems decomposition

The PMSG model given by Eqs. (3.6)−(3.8)can be re-arranged as follows:

$$\sum_e : v_e = \begin{bmatrix} v_{dq} \\ -\omega_m \end{bmatrix} \rightarrow Y_e = \begin{bmatrix} i_{dq} \\ T_m \end{bmatrix} \tag{3.9}$$

$$\sum_m : v_m = (-T_e + T_m) \rightarrow Y_e = -\omega_m = \frac{(-T_e + T_m)}{Js + f_{fv}} \tag{3.10}$$

The Lemma 3.1 is formulated based on the above Eqs. (3.9) and (3.10), as expressed below:

Lemma 3.1. *the model* (3.6)−(3.8) *is decomposable into two subsystems interconnected with passive feedback, that is electrical dynamic* $\sum_e$ and mechanical dynamic $\sum_m$ based on the above conditions.

Proof. from $\sum_e$, the total energy H_e is given below:

$$H_e = \frac{1}{2}\, i_{dq}^T L_{dq}\, i_{dq} + \psi_{dq}^T\, i_{dq} \tag{3.11}$$

Time derivative along Eq. (3.6) of H_e, gives:

$$\dot{H}_e = -\, i_{dq}^T R_{dq}\, i_{dq} + Y_e^T v_e + \frac{d}{dt}\left(\psi_{dq}^T i_{dq}\right) \tag{3.12}$$

Integrating Eq. (3.12) along $\begin{bmatrix} 0 & T_e \end{bmatrix}$, yields:

$$\underbrace{H_e(T_e) - H_e(0)}_{\text{Stored Energy}} = \underbrace{-\int_0^{T_e} i_{dq}^T R_{dq}\, i_{dq}\, d\tau}_{\text{Dissipated Energy}} + \underbrace{\int_0^{T_e} Y_e^T v_e d\tau + \left[\psi_{dq}^T\, i_{dq}\right]_0^{T_e}}_{\text{Supplied Energy}} \tag{3.13}$$

where, $H_e(0)$ denotes the initial stored energy and $H_e(T_e) \geq 0$. Integrating Eq. (3.13), yields the following dissipation inequality:

$$\int_0^{T_e} Y_e^T v_e d\tau \geq \lambda_{\min}\{R_{dq}\} \int_0^{T_e} i_{dq}{}^2 d\tau - \left(H_e(0) + \left[\psi_{dq}^T\, i_{dq}\right]_0^{T_e}\right) \tag{3.14}$$

where $\| \cdot \|$ denotes the Euclidian vector norm.

Then, it is deduced from Eq. (3.14) that $\sum_e$ is passive. Thus $F_m(s)$ is given as below:

$$F_m(s) = \frac{Y_m(s)}{V_m(s)} = \frac{1}{Js + f_{fv}} \tag{3.15}$$

Since $F_m(s)$ is a strictly positive, the mechanical dynamic $\sum_m$ is passive. Then, the PMSG model is decomposable into $\sum_e$ and $\sum_m$.

3.3.2 Permanent magnet synchronous generator passivity property

Lemma 3.2. *The model* (3.6)−(3.8) *is passive, when* $Y = \left[v_{dq}^T, T_e\right]^T$ and $X = \left[i_{dq}^T, \omega_m\right]^T$ are chosen as the PMSG outputs and inputs, respectively.

Proof. First, the PMSG Hamiltonian H_m is defined as:

$$H_m\left(i_{dq}, \omega_m\right) = \underbrace{\frac{1}{2} i_{dq}^T L_{dq}\, i_{dq} + \psi_{dq}^T\, i_{dq}}_{\text{Electrical Energy}} + \underbrace{\frac{1}{2} J\omega_m^2}_{\text{Mecanical Energy}} \tag{3.16}$$

Derivative along Eqs. (3.6)−(3.8) of H_m, gives:

$$\frac{dH_m\left(i_{\alpha\beta}, \omega_m\right)}{dt} = -\frac{d\left(i_{\alpha\beta}^T\, R i_{dq}\right)}{dt} + y^T\nu + \frac{d}{dt}\left(\psi_{dq}^T\, i_{dq}\right) \tag{3.17}$$

Where: $R = \text{diag}\left\{R_{dq}, f_{fv}\right\}$. Integrating Eq. (3.17) along $\left[0 \quad T_m\right]$, gives:

$$\underbrace{H_m(T_m) - H_m(0)}_{\text{Stored Energy}} = \underbrace{-\int_0^{T_m} i_{dq}^T\, R i_{dq} d\tau}_{\text{Dissipated Energy}} + \underbrace{\int_0^{T_m} y^T \nu d\tau + \left[\psi_{dq}^T\, i_{dq}\right]_0^{T_m}}_{\text{Supplied Energy}} \tag{3.18}$$

Where $H_m(0)$ is the stored initial energy and $H_m(T_m) \geq 0$. Integrating (3.18), yields:

$$\int_0^{T_m} y^T \nu d\tau \geq \lambda_{min}\{R\} \int_0^{T_m} {i_{dq}}^2 d\tau - \left(H_m(0) + \left[\psi_{dq}^T\, i_{dq}\right]_0^{T_m}\right) \tag{3.19}$$

Then, relationship M is passive, which is the same for the PMSG.

3.3.3 Workless forces identification

From the model Eqs. (3.6)–(3.8), the following compact form can be deduced:

$$d\frac{di_{dq}}{dt} + \mathrm{WX} + \mathrm{RX} = \mathrm{Mv}_{dq} + \Gamma \tag{3.20}$$

where, $D = \mathrm{diag}\{L_{dq}, J\}, m = [i_2, 0_{1\times 2}]^T, W = \left[\frac{d}{dt}\psi_{dq}^T\omega_m, -i_{dq}\frac{d}{dt}\psi_{dq}\right]^T$, and $\Gamma = [0_{2\times 1}, -T_m]^T$. Based on the PMSG model (3.20) and its passivity property, the "workless forces" are deduced as given below:

$$F = \begin{bmatrix} 0_{2\times 2} & \dfrac{d}{dt}\psi_{dq} \\ -\dfrac{d}{dt}\psi_{dq}^T & 0_{1\times 1} \end{bmatrix} \tag{3.21}$$

As F satisfy:

$$F = -F^T \tag{3.22}$$

Remark 3.1. from Sections 3.1, 3.2, and 3.3, the necessary conditions to apply the PBSC on the PMSG (see Section 3) have been analytically verified and satisfied.

3.3.4 Speed-controlled dq model of the permanent magnet synchronous generator

In order to design the control strategy, based on the passivity concept, the state-space model of the PMSG is written in the following form [15]:

$$\mathscr{L}\dot{X} = \mathscr{J}(u)X - \mathscr{R}X + \mathscr{B}u + \xi \tag{3.23}$$

Where $\mathscr{L} = \begin{bmatrix} L_d & 0 & 0 \\ 0 & L_q & 0 \\ 0 & 0 & \dfrac{2J}{3p^2} \end{bmatrix}$ is a constant symmetric positive definite matrix, $\mathscr{R} = \begin{bmatrix} R_s & 0 & 0 \\ 0 & R_s & 0 \\ 0 & 0 & \dfrac{2f_{fv}}{3p^2} \end{bmatrix}$ is the losses matrix,

$\mathcal{J}(u) = \begin{bmatrix} 0 & L_d\omega_m & 0 \\ -L_d\omega_m & 0 & -\phi_f \\ 0 & \phi_f & 0 \end{bmatrix}$ is a skew symmetric matrix, $\xi = \begin{bmatrix} 0 \\ 0 \\ -\dfrac{2T_m}{3\text{p}} \end{bmatrix}$ is the disturbance vector, $\mathcal{B} = \begin{bmatrix} 1 & 0 \\ 0 & 1 \\ 0 & 0 \end{bmatrix}$, and u denote the input-matrix which is the controller law defined as

$$u = \begin{bmatrix} v_d \\ v_q \end{bmatrix} \tag{3.24}$$

To guarantee that the defined state model (3.23) of the PMSG, is the system that can be controlled by passivity, the matrices $\mathcal{J}(u)$ and $\mathcal{R}$ must satisfy the conditions $\mathcal{J}^T(u) = -\mathcal{J}(u)$ and $\mathcal{R}^T = \mathcal{R}$.

By considering that the aforementioned conditions are satisfied, then the reference voltage vector $u^* = \begin{bmatrix} v_d^* & v_q^* \end{bmatrix}^T$ is considered as the desired input and $X^* = \begin{bmatrix} i_q^* & i_q^* & \omega_m^* \end{bmatrix}^T$ is the desired state variables. This yields the desired dynamic state model of the PMSG given below:

$$\mathcal{L}\dot{X}^* = \mathcal{J}(\text{u})X^* - \mathcal{R}X^* + \mathcal{B}u^* + \xi \tag{3.25}$$

The desired control input u^* is designed by the passivity approach, using the previous simplified model (3.25).

3.4 Passivity-based speed controller computation

The differences between u and u^*, X and X^* representing the voltage tracking error and the state variables tracking error, respectively, as described by:

$$e_u = \begin{bmatrix} e_{ud} \\ e_{uq} \end{bmatrix} = u - u^* \tag{3.26}$$

$$e = \begin{bmatrix} e_d \\ e_q \end{bmatrix} = X - X^* \tag{3.27}$$

Substituting Eq. (3.23) in Eq. (3.25), yields

$$\mathcal{L}\dot{e} = [\mathcal{J}(u) - \mathcal{J}(u^*)]X - \mathcal{R}e + \mathcal{B}e_u + \mathcal{J}(u^*)e \tag{3.28}$$

Principally, it is a linear control for the nonlinear dynamical of the PMSG. By using Taylor series, $\mathcal{J}(u)$ can be linearized as follows:

$$\mathcal{J}(u) = \mathcal{J}(u^*) + \frac{d\mathcal{J}(u)}{du}|_{u^*} e_u \tag{3.29}$$

where $\frac{d\mathcal{J}(u)}{du}|_{u^*} e_u = 0$ because $\mathcal{J}(u)$ is a constant matrix.

Substituting Eq. (3.29) in Eq. (3.28), yields

$$\mathcal{L}\dot{e} = -\mathcal{R}e + \mathcal{B}e_u + \mathcal{J}(u^*)e \tag{3.30}$$

The aim is to ensure the convergence to zero of the error vectors e_u and the stability property of Eq. (3.30), by finding the desired input u^*. Using the Lyapunov theory, the following energy function of the closed-loop system is defined by:

$$v(\mathrm{e}) = 0.5\mathrm{e}^T \mathcal{L} \mathrm{e} \tag{3.31}$$

Taking the time derivative of $v(\mathrm{e})$ along trajectory Eq. (3.30), gives

$$\dot{v}(\mathrm{e}) = -\mathrm{e}^T \mathcal{R} \mathrm{e} + \mathrm{e}^T \mathcal{B} e_u \tag{3.32}$$

The term $\mathrm{e}^T \mathcal{J}(u^*)e$ does not appear on the right-side of Eq. (3.32), because of $\mathrm{e}^T \mathcal{J}(u^*)$ which is nonsymmetrical. By considering $e_u = -k_e \mathcal{B}^T e$, the Eq. (3.32) become:

$$\dot{v}(\mathrm{e}) = -\mathrm{e}^T \left(\mathcal{R} + \mathcal{B} k_e \mathcal{B}^T\right) e \tag{3.33}$$

where, $k_e = \begin{bmatrix} k_1 & 0 \\ 0 & k_2 \end{bmatrix}$ with $k_1 > 0$ and $k_2 > 0$.

Taking $\left(\mathcal{R} + \mathcal{B} k_e \mathcal{B}^T\right) \geq 0$ with the aim to make the energy function $\dot{v}(\mathrm{e})$ negative defined, which guarantee the stability of the Eq. (3.30) and assuming that $e_u = u - u^* = -k_e \mathcal{B}^T e$, which gives:

$$e_u = -\begin{bmatrix} k_1 & 0 \\ 0 & k_2 \end{bmatrix} \begin{bmatrix} 1 & 0 & 0 \\ 0 & 1 & 0 \end{bmatrix} \begin{bmatrix} i_d - i_d^* \\ i_q - i_q^* \\ \omega_m - \omega_m^* \end{bmatrix} = -\begin{bmatrix} k_1\left(i_d - i_d^*\right) \\ k_2\left(i_q - i_q^*\right) \end{bmatrix} \tag{3.34}$$

Then, the control signals $u = \begin{bmatrix} v_d & v_q \end{bmatrix}^T$ is deduced, which ensure the convergence of the voltage tracking error e_u. Then, the control voltage vector is given by:

$$u = \begin{bmatrix} v_d \\ v_q \end{bmatrix} = \begin{bmatrix} v_d^* - k_1\left(i_d - i_d^*\right) \\ v_q^* - k_2\left(i_q - i_q^*\right) \end{bmatrix} \tag{3.35}$$

The voltage controller u, consists of two parts: the desired reference vector u^* and the damping term to make the closed-loop system strictly passive. However, it is well known that fixed gains are very sensitive when the system is exposed to parameter uncertainties and external disturbances. Thus to overcome this problem and to compute un optimal controller, a fuzzy controller is introduced as a supervisor to compute the damping gains k_1 and k_2 to overcome the problem caused by parameter uncertainties which makes the proposed PBSC intelligent (see Fig. 3.3). The current error $e_i = (i_d - i_d^*)$ and its derivative are taken as the inputs of the fuzzy supervisor. The selected fuzzy control design process consists of: fuzzification of the inputs, formulation of the rules, and finally defuzzification of the output. Triangular and trapezoidal types symmetrical and uniformly distributed are used to select the membership functions as given in Fig. 3.4. Lee and Yubazaki [21] and [22] method is used of partitioning these functions which consists on sharing the same parameter by several membership functions. The significant reduced of the parameter numbers of the membership functions is the advantage this method. The inputs-outputs linguistic variables corresponding to of the fuzzy block are tabulated in Table 3.1, which are defined as: Positive Small (PS), Positive Big (PB), Zero (Z), Negative Big (NB), and Negative Small (NS). A Max-Min fuzzy inference is used for the decision-making where the center of gravity defuzzification method is used to calculate the crisp outputs [21].

Remark 3.2. As shown in Fig. 3.2, the proposed method ensures the current and speed tracking errors convergence. The damping term

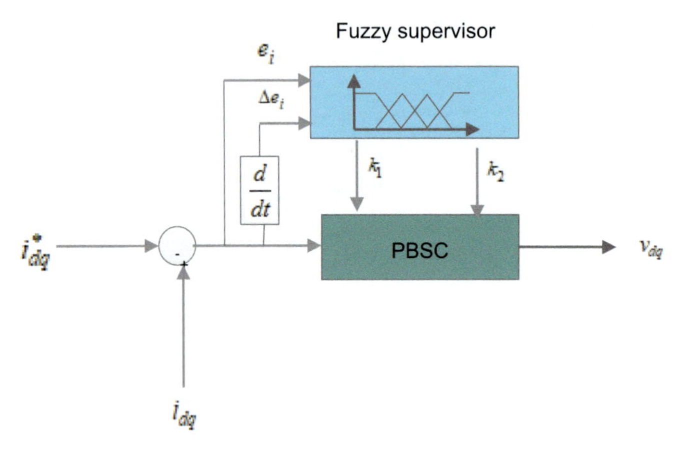

Figure 3.3 Damping gains with the fuzzy supervisor.

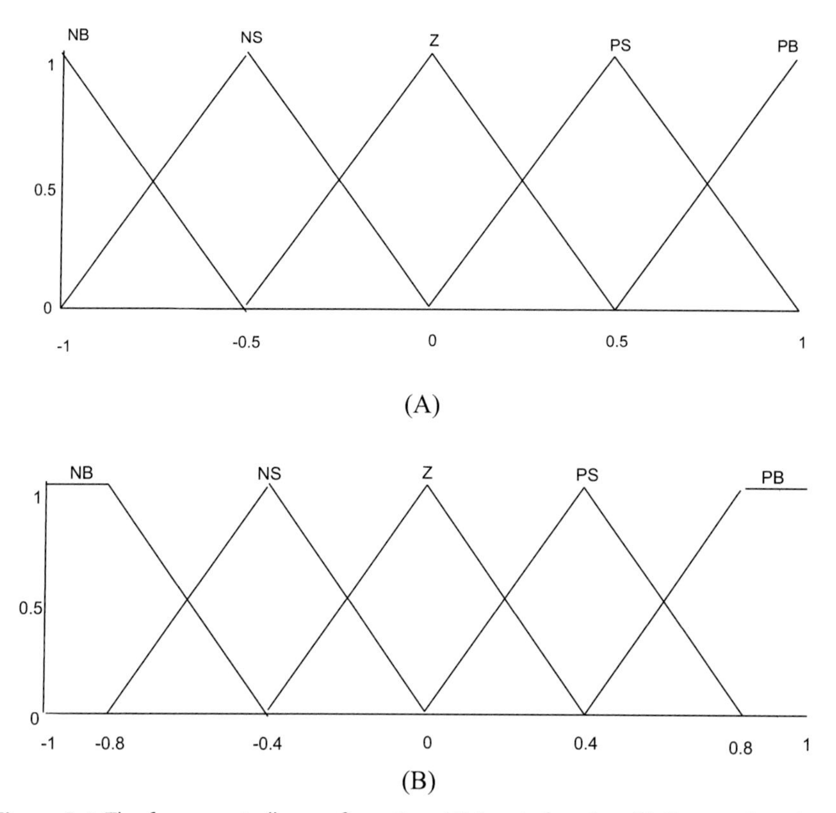

Figure 3.4 The fuzzy controller configuration. (A) Inputs function (B) Outputs function.

Table 3.1 Fuzzy logic rules.

$\Delta\varepsilon_i$ ε_i	**NB**	**NS**	**Z**	**PS**	**PB**
NB	NB	NB	NS	NS	Z
NS	NB	NB	NS	Z	PS
Z	NS	NS	Z	PS	PS
PS	NS	Z	PS	PB	PB
PB	Z	PS	PS	PB	PB

$"k_e\mathscr{B}^T e"$ ensures convergence of the voltage tracking error $"e_u"$ [Eq. (3.34)]. The condition $\left(\mathscr{R} + \mathscr{B}k_e\mathscr{B}^T\right) \geq 0$ ensures a negative time derivative of the function $\dot{v}(e)$ Eq. (3.33), thus the convergence of the voltage tracking error $"e_u"$, and then, the equality between the voltage u which is the controller law and the desired u^* Eq. (3.35). Therefore the convergence of the voltage tracking error $"u - u"$ should be ensured.

3.4.1 Desired voltage and desired current computation

In order to design the control strategy, based on the passivity concept, the state-space model of the PMSG is written in the following form [15]:

The voltage reference u^* and the desired stator currents $i_{dq}^* = \begin{bmatrix} i_d^* & i_q^* \end{bmatrix}^T$ are computed from the system equilibrium point $\left(\dot{X} = 0\right)$ described by Eq. (3.34):

$$\begin{bmatrix} L_d \omega_m^* i_q^* - R_s i_d^* + v_d^* \\ -\omega_m^* \left(L_d i_q^* + \phi_f\right) - R_s i_d^* + v_q^* \\ \phi_f i_q^* - \dfrac{2}{3\mathrm{p}} \left(T_m + \dfrac{f_{fv}}{\mathrm{p}} \omega_m^*\right) \end{bmatrix} = \begin{bmatrix} 0 \\ 0 \\ 0 \end{bmatrix} \tag{3.36}$$

Knowing that the PMSG operate under maximum torque when the desired direct current $i_d^* = 0$. Under this condition, and using Eq. (3.36), it yields

$$u^* = \begin{bmatrix} v_d^* \\ v_q^* \end{bmatrix} = \begin{bmatrix} -L_d \omega_m^* i_q^* \\ \omega_m^* \left(L_d i_q^* + \phi_f\right) \end{bmatrix} \tag{3.37}$$

$$i_q^* = \frac{2}{3\mathrm{p}\phi_f} \left(T_m + \frac{f_{fv}}{\mathrm{p}} \omega_m^*\right) \tag{3.38}$$

3.5 Grid-side converter control

To regulate and transmit to electrical energy produced by the PMSG to the grid through the GSC, a classical method is selected which consists on PI strategy depicted in Fig. 3.5. The GSC mathematical model is expressed as follows [1,19]:

$$\begin{bmatrix} v_d \\ v_q \end{bmatrix} = R_f \begin{bmatrix} i_{df} \\ i_{qf} \end{bmatrix} + \begin{bmatrix} L_f \dfrac{di_{df}}{dt} - \omega L_f i_{qf} \\ L_f \dfrac{di_{qf}}{dt} + \omega L_f i_{df} \end{bmatrix} + \begin{bmatrix} v_{gd} \\ v_{gq} \end{bmatrix} \tag{3.39}$$

Where, v_{gd} and v_{gq} are the grid voltages, i_{df} and i_{qf} are the grid currents, L_f is the filter inductance, R_f represents the filter resistance,

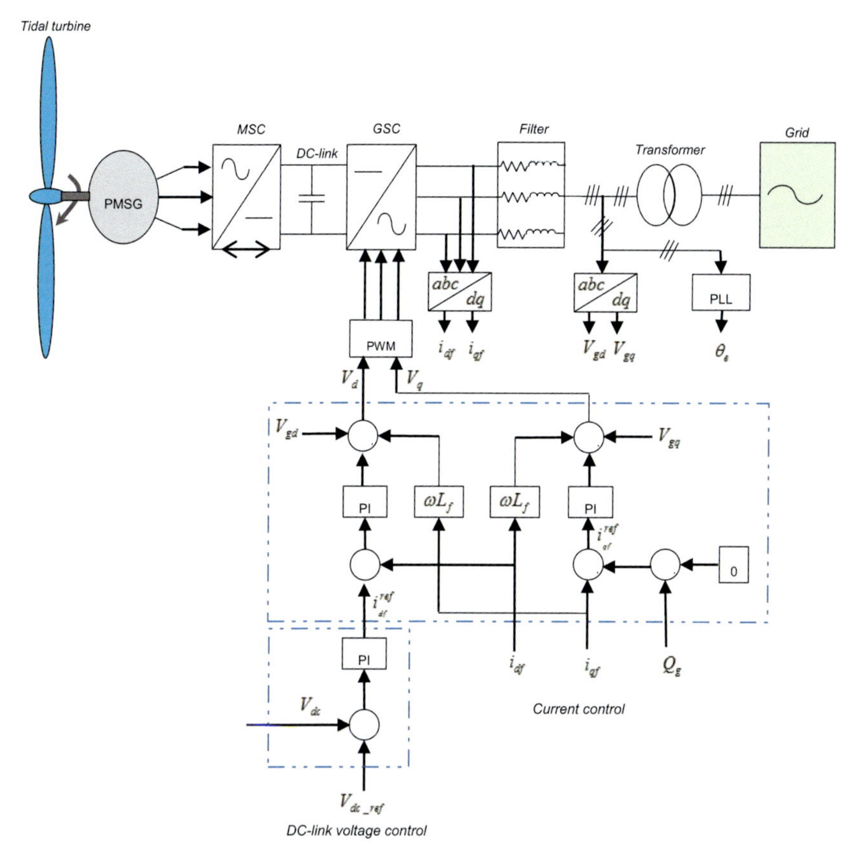

Figure 3.5 Grid-side control strategy.

ω denotes the network angular frequency, and v_d and v_q denotes the inverter voltages. The DC link voltage mathematical model is formulated below [1]:

$$C\frac{dv_{dc}}{dt} = \frac{3}{2}\frac{v_{gd}}{v_{dc}}i_{df} + i_{dc} \tag{3.40}$$

where, C is the DC-link capacitance, i_{dc} is the line current, and v_{dc} is the DC-link voltage. The mathematical model of PI current loop is formulated as:

$$\begin{cases} v_{gd}^{\mathrm{PI}} = k_{gp}^{d}\left(i_{df}^{ref} - i_{df}\right) - k_{gi}^{d}\int_0^t \left(i_{df}^{ref} - i_{df}\right)d\tau \\ v_{gq}^{\mathrm{PI}} = k_{gp}^{q}\left(i_{qf}^{ref} - i_{qf}\right) - k_{gi}^{q}\int_0^t \left(i_{qf}^{ref} - i_{qf}\right)d\tau \end{cases} \tag{3.41}$$

where, $k_{gp}^{d} > 0$, $k_{gi}^{d} > 0$, $k_{gp}^{q} > 0$, $k_{gi}^{q} > 0$. The mathematical model of DC-link PI loop is formulated as:

$$i_{qf}^{ref} = k_{dcp}\left(v_{dc\_ref} - v_{dc}\right) - k_{dci}\int_{0}^{t}\left(v_{dc\_ref} - v_{dc}\right)d\tau \qquad (3.42)$$

where, $k_{dcp} > 0$ and $k_{dci} > 0$. Finally, the mathematical model of the active and reactive powers is formulated as below:

$$\begin{cases} P_g = \frac{3}{2} v_{gd}\, i_{df} \\ q_g = \frac{3}{2} v_{gd}\, i_{qf} \end{cases} \qquad (3.43)$$

3.6 Simulation and experimental results

The numerical investigation is performed on the conversion by using Matlab/Simulink showed by Fig. 3.1. The parameter values of the closed-loop are given in Table 3.2. 1150 V and zero values are chosen as the reference command for v_{dc} and q_g, respectively. The initial values of the parameters are given as: $[\omega_m(0), i_{dq}(0)] = [0, 0, 0]$, $v_{dc}\ (0) = 0$ and $i_{dqf}(0) = [0, 0]$. Using the pole location method, the GSC current PI gains are given as: $k_{gp}^{d} = k_{gp}^{q} = 9$ and $k_{gi}^{d} = k_{gi}^{q} = 200$. The DC-link PI gains are:

Table 3.2 System parameters.

PMSG parameter	Value
Tidal density (ρ)	1024 kg/m^2
Tidal turbine radius (R)	10 m
Stator inductance (L_{dq})	0.3 mH
Stator resistance (R_s)	0.006 Ω
Stator inductance (L_{dq})	0.3 mH
Pole pairs number (p)	48
Flux linkage (ϕ_f)	1.48 Wb
Total inertia (J)	35,000 kg.m^2
DC-link voltage (V_{dc})	1150 V
DC-link capacitor (C)	2.9 F
Grid-filter resistance (R_f)	0.3 pu
Grid-filter inductance (L_f)	0.3 pu

$k_{dcp} = 5$ and $k_{dci} = 500$. The investigated strategy is compared with the fuzzy passivity-based linear feedback current control (FPBLFC) proposed in Belkhier and Achour [14], the passivity-based current control (PBCC) proposed in Achour et al. [15], and the high-order sliding mode control (HSMC) [7]. The proposed method is tested under two scenarios: First, the proposed controller is tested with initial parameter values of the PMSG and compared with the benchmark controls. The second task, deal with the robustness analysis of this proposed method due to parameter changes.

3.6.1 Performance analysis under fixed parameters

Fig. 3.6 depict the profile of the tidal speed imposed to the conversion system. Fig. 3.7 depicts the DC voltage response due to proposed FS-PBSC, FPBLFC, PBCC, and HSMC controls. Given the showed responses and Table 3.3, a transient undershoots of -0.002 and overshoot of $+0.002$ is observed with the proposed FS-PBSC, a transient undershoots of -0.02, -0.03, and -0.2 are observed with FPBLFC, PBCC, and HSMC methods, respectively and transient overshoots of $+0.02$, $+0.03$, and $+0.2$. From the tracing response of the DC-bus given by Fig. 3.7, it can be seen that the DC voltage error $[\varepsilon\ (v_{dc})]$ is extremely reduced in case of the proposed FS-PBSC. Figs. 3.8 and 3.9 illustrate the tracking error $[\varepsilon\ (q_g)]$ of q_g due to the investigated FS-PBSC, FPBLFC, PBCC, and HSMC where transient under and overshoots of $-1.5e-5$, $-4e-5$, $-5e-5$, $-7e-5$ and $+1.5e-5$, $+4e-5$, $+5e-5$, $+7e-5$ are observed, respectively. However, lowest under and overshoot are illustrated by the proposed FS-PBSC

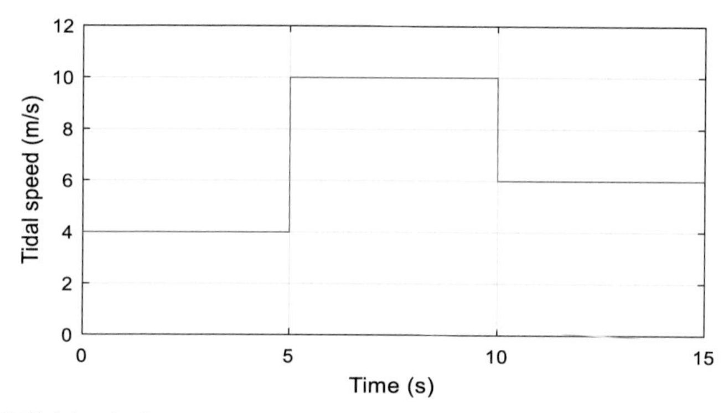

Figure 3.6 Tidal velocity.

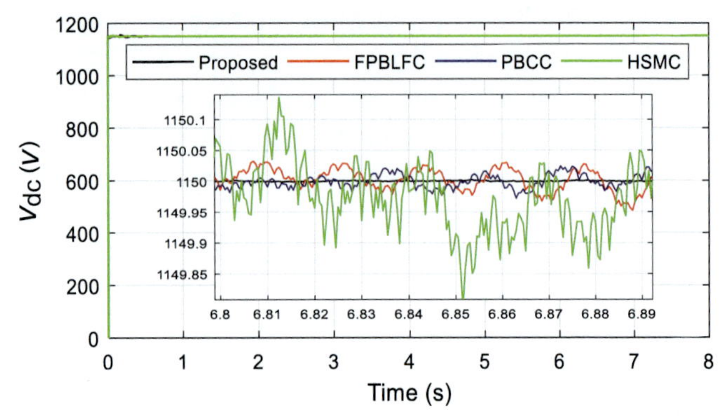

Figure 3.7 Zoom on DC-link voltage response.

Table 3.3 Performance analysis under initial parameter values.

Control	Proposed	Fuzzy passivity-based linear feedback current	Passivity-based current control	High-order sliding mode control
Variation	**R_s and J**	***Rs* and *J***	**R_s and J**	**R_s and J**
V_{dc}(V)	$\pm$ 1150.002	$\pm$ 1150.02	$\pm$ 1150.03	$\pm$ 11502
Q_g(MW)	$\pm$ 1.5^{e-5}	$\pm$ 4^{e-5}	$\pm$ 5^{e-5}	$\pm$ 7^{e-5}
$\varepsilon(V_{dc})$	$\pm$ 0.002	$\pm$ 0.02	$\pm$ 0.03	$\pm$ 02
$\varepsilon(Q_g)$	$\pm$ 0.000015	$\pm$ 0.00004	$\pm$ 0.00004	$\pm$ 0.00007

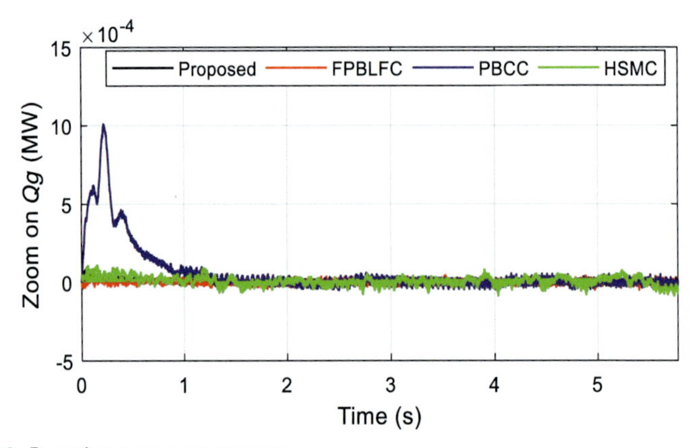

Figure 3.8 Reactive power response.

according to Table 3.3. Moreover, lowest steady-state error and better convergence criterion is shown by the proposed FS-PBSC (0.3e−3s) over the FPBLFC (1e−3s), PBCC (1.2e−3s), and HSMC

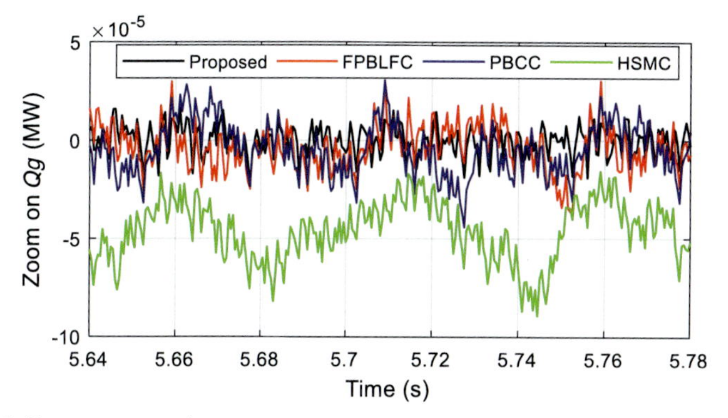

Figure 3.9 Zoom on reactive power.

(2e−3s) as depicted by. Generally, the proposed FS-PBSC ensured fast convergence, high efficiency and lowest tracking errors in comparison to the tested benchmark nonlinear strategies.

3.6.2 Robustness analysis

In the present section, the system is supposed to work under parameter disturbances. Thus a change of $+50\%$ in R_s, a change of $+100\%$ in J, and a simultaneous change of $+50\%$ in R_s and $+100\%$ in J are imposed on the closed-loop, respectively. Fig. 3.10 shows the DC-link voltage response due to the proposed candidate in case of $+50\%$ in R_s, where the same voltage error ε (V_{dc}) response and tracking error equal to zero are almost recorded, that is ± 0.002 which is the same as in subsection 5.1. Similarity tracking response of q_g for the diturbances of $+50$ R_s is also observed in Fig. 3.11, that is $\pm 2e-5$. Fig. 3.12 shows the DC-link voltage response due to the proposed FS-PBSC in case of variation of $+100\%$ in J, here also, no changes in the voltage error ε (V_{dc}) response are recorded, that is ± 0.002. Fig. 3.13 shows the q_g response for the diturbances of $+100$ J, one can see that this disturbance does not influence the dynamic of the system, that is $\pm 1.5^{e-5}$. Simultaneous change of $+50\%$ in R_s and $+100\%$ in J have no effect on both v_{dc} and q_g according to Figs. 3.14 and 3.15. From the given results, the proposed FS-PBSC shows the same behavior due to both fixed and varied parameter values, which clearly demonstrated the robustness of the proposed strategy against parameter uncertainties.

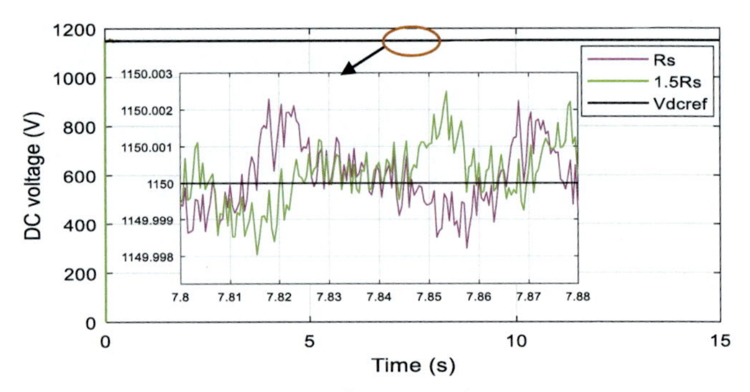

Figure 3.10 DC-link voltage response of $+50\%$ of R_s.

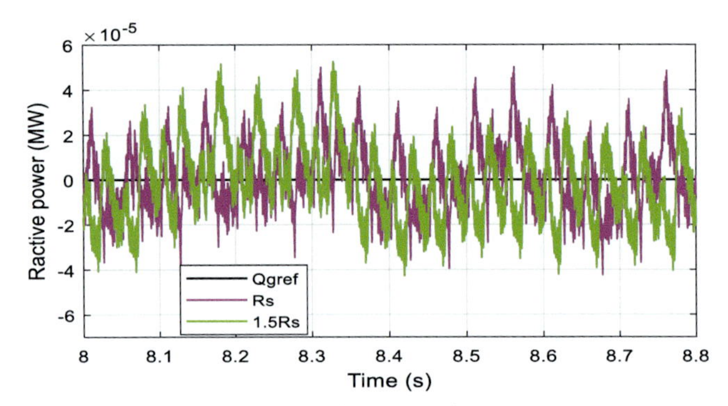

Figure 3.11 Reactive power response of $+50\%$ of R_s.

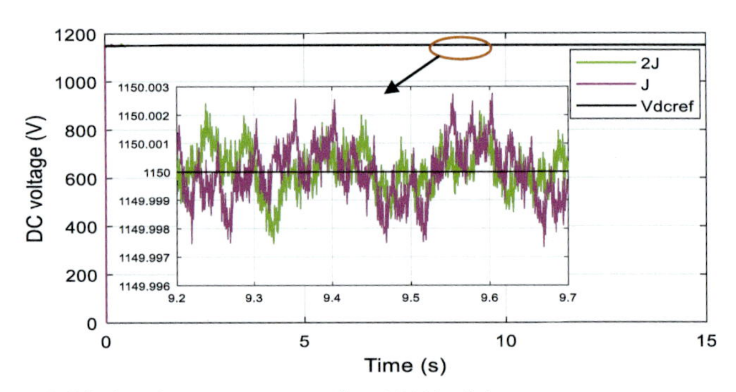

Figure 3.12 DC-link voltage response of $+100\%$ of J.

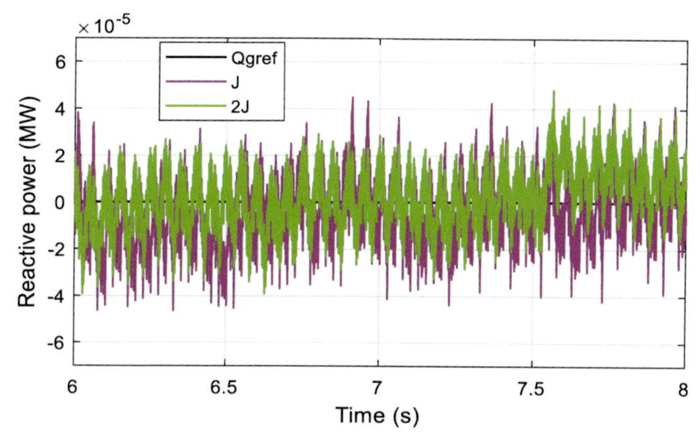

Figure 3.13 Reactive power response of + 100% of J.

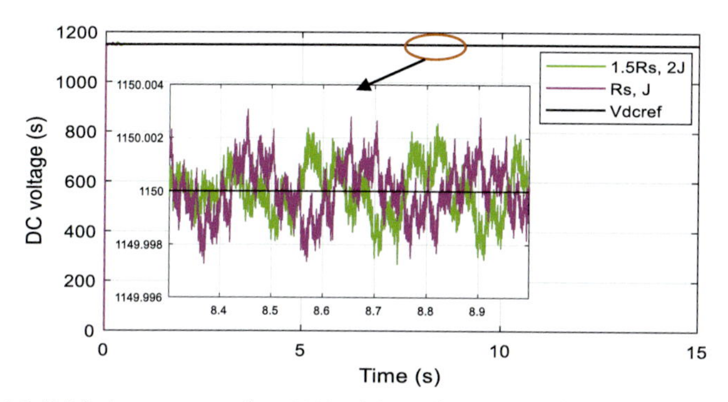

Figure 3.14 DC-link response of + 50% of R_s and + 100% of J.

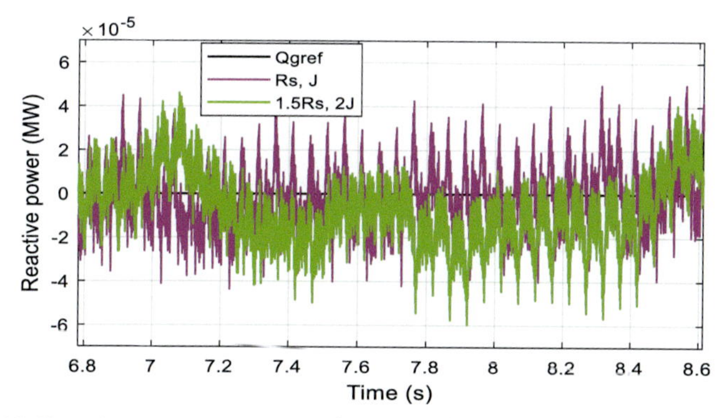

Figure 3.15 Reactive power response of + 50% of R_s and + 100% of J.

3.7 Conclusion

New fuzzy supervisor-based speed controller for a PMSG in tidal turbine conversion system is proposed. The proposed strategy is adopted to extract the maximum power from the tidal energy where the entire dynamics of the PMSG is considered when designing the control law. The fuzzy controller is selected to guarantee the overall-rated speed operation of the PMSG and then, compute optimal damping gains. Dynamic simulations of the studied system under parameter changes have taken special attention and the results have been compared to nonlinear control methods, which show a quick track of the reactive power and the DC-link voltage to their references over the compared controls. Also, it is observed that the closed-loop operate at the maximum power and integrate an efficient electrical power to the grid. The proposed candidate offers high robustness, fast speed convergence, and high efficiency over the other benchmark nonlinear strategies.

References

[1] Z. Zhou, M. Benbouzid, J.F. Charpentier, F. Scuiller, T. Tang, Developments in large marine current turbine technologies – a review, Renew. Sustain. Energy Rev. 71 (2017) 852–858.

[2] E. Mohammadi, R. Fadaeinedjad, H.R. Nadji, Design, electromechanical simulation, and control of a variable speed stall-regulated PMSG-based wind turbine, Int. J. Green Energy 16 (12) (2019) 890–900.

[3] S.W. Wang, Adaptive fuzzy robust control of PMSM with smooth inverse based dead-zone compensation, Int. J. Control Autom. Syst. 14 (2) (2020) 378–388.

[4] Y.-J. Gu, X.-X. Yin, H.-W. Liu, W. Li, Y.-G. Lin, Fuzzy terminal sliding mode control for extracting maximum marine current energy, Energy 90 (1) (2015) 258–265.

[5] A.M. Othman, Enhancement of tidal generators by superconducting energy storage Jaya-based sliding-mode controller, Int. J. Energy Res. 44 (14) (2020) 11658–11675.

[6] X. Yin, X. Zhao, ADV Preview based nonlinear predictive control for maximizing power generation of a tidal turbine with hydrostatic transmission, IEEE Trans. Energy Convers. 34 (4) (2019) 1781–1791.

[7] Z. Zhou, F. Scuiller, J.F. Charpentier, M.E.H. Benbouzid, T. Tang, Power smoothing control in a grid-connected marine current turbine system for compensating swell effect, IEEE Trans. Sustain. Energy 4 (3) (2013) 816–826.

[8] Z. Zhou, B.S. Elghali, M.E.H. Benbouzid, Y. Amirat, E. Elbouchikhi, G. Feld, Tidal stream turbine control: an active disturbance rejection control approach, Ocean. Eng. 202 (2020) 107190.

[9] P.C. Sahu, R. Baliarsingh, R.C. Prusty, S. Panda, Novel DQN optimized tilt fuzzy cascade controller for frequency stability of a tidal energy based AC Microgrid, Int. J. Ambient. Energy (2020). Available from: https://doi.org/10.1080/01430750.2020.1839553.

[10] S. Toumi, E. Elbouchikhi, Y. Amirat, M. Benbouzid, G. Feld, Magnet failure-resilient control of a direct-drive tidal turbine, Ocean. Eng. 187 (2019) 106207.
[11] R. Gaamouche, A. Redouane, I. El harraki, B. Belhorma, A. El hasnaoui, Optimal feedback control of nonlinear variable-speed marine current turbine using a two-mass model, J. Mar. Sci. Appl. 19 (2020) 83−95.
[12] S.H. Moon, B.G. Park, J.W. Kim, J.M. Kim, Maximum power-point tracking control using perturb and observe algorithm for tidal current generation system, Int. J. Precis. Eng. Manuf.-Green Technol. 7 (2020) 849−858.
[13] B. Yang, Q.H. Wu, L. Tiang, J.S. Smith, Adaptive passivity-based control of a TCSC for the power system damping improvement of a PMSG based offshore wind farm, in: Proceedings of the IEEE International Conference on Renewable Energy Research and Applications ICRERA, 2013, Madrid, Spain, 1−5.
[14] Y. Belkhier, A.Y. Achour, Fuzzy passivity-based linear feedback current controller approach for PMSG-based tidal turbine, Ocean. Eng. 218 (2020) 108156. Available from: https://doi.org/10.1016/j.oceaneng.2020.108156.
[15] A.Y. Achour, B. Mendil, S. Bacha, I. Munteanu, Passivity-based current controller design for a permanent-magnet synchronous motor, ISA Trans. 48 (3) (2009) 336−346.
[16] B. Yang, H. Yu, Y. Zhang, J. Chen, Y. Sang, L. Jing, Passivity-based sliding-mode control design for optimal power extraction of a PMSG based variable speed wind turbine, Renew. Energy 119 (2018) 577−589.
[17] R. Subramaniam, Y.H. Joo, Passivity-based fuzzy ISMC for wind energy conversion systems with PMSG, IEEE Trans. Syst., Man, Cyber. Syst. 51 (4) (2021) 2212−2220. Available from: https://doi.org/10.1109/TSMC.2019.2930743.
[18] B. Yang, T. Yu, H. Shu, D. Qiu, Y. Zhang, P. Cao, et al., Passivity-based linear feedback control of permanent magnetic synchronous generator-based wind energy conversion system: design and analysis, IET Renew. Power Gener. 12 (9) (2018) 981−991.
[19] N. Khefifi, A. Houari, M. Machmoum, M. Ghanes, M. Ait-Ahmed, Control of grid forming inverter based on robust IDA-PBC for power quality enhancement, Sustain. Energy Grids Netw. 20 (2019) 100276.
[20] Y. Belkhier, A.Y. Achour, An intelligent passivity-based backstepping approach for optimal control for grid-connecting permanent magnet synchronous generator-based tidal conversion system, Int. J. Energy Res. 45 (4) (2021) 5433−5448. Available from: https://doi.org/10.1002/er.6171.
[21] M.A. Lee, H. Takagi, Dynamic control of genetic algorithms using fuzzy logic techniques, in: Proceedings of International Conference on Genetic Algorithms, 1993, San Mateo, CA, pp. 76−83.
[22] N. Yubazaki, M. Otami, T. Ashid, K. Hirota, Dynamic fuzzy control method and its application to positioning of induction motor, Proc. 1995 IEEE Int. Conf. Fuzzy Syst. 3 (1995) 1095−1102.

CHAPTER FOUR

An intelligent energy management system of hybrid solar/wind/battery power sources integrated in smart DC microgrid for smart university

Youcef Belkhier[1], Mohamed Dahman Alshehri[2], Rabindra Nath Shaw[3] and Ankush Ghosh[4]

[1]Department of Computer Science, Faculty of Electrical Engineering, Czech Technical University in Prague, Prague, The Czech Republic
[2]Department of Computer Science, College of Computers and Information, Taif University, Taif, Saudi Arabia
[3]Department of International Relations, Bharath Institute of Higher Education and Research (Deemed to be University), Chennai, India
[4]School of Engineering and Applied Sciences, The Neotia University, Sarisha, India

4.1 Introduction

In order to satisfy future demands, renewable energy sources (RESs) need to be integrated into the power system due to the fast use and pollution of fossil fuels. Solar energy, geothermal energy, ocean energy, wind energy, bioenergy, and other RESs have been deployed in many locations around the world, particularly in rural areas [1,2]. The most common RESs are wind and solar energy. Small-scale off-grid (microgrid) systems are established in remote locations rather than establishing transmission lines to transfer power from generation units to loads. A microgrid system is a tiny system that mostly uses solar and wind energy. Increased nonrenewable energy supplies and storage of energy have also increased in order to guarantee the permanent and stable power supply due to instability, intermission, and the high cost of solar and wind power systems. Hybrid renewable energy systems are created when RESs are combined with other energy sources [3,4]. The demand for energy by consumers is generally not evenly distributed over time and problems of the phasing of energy produced versus energy consumed arise. The stability of the grid

Applications of AI and IOT in Renewable Energy.
DOI: https://doi.org/10.1016/B978-0-323-91699-8.00004-8

depends on the balance between production and consumption [3]. The increase in the penetration rate of renewable energies will therefore be conditioned by their participation in these different services, which will be favored by the association with these clean energy sources, of electrical energy storage systems (ESSs) [4]. Storage is therefore the key to the penetration of these energies in the electricity grid. Not only does it provide a technical solution for the grid operator to ensure a real-time balance of production and consumption, but it also enables the best possible use of renewable resources by avoiding load shedding in the event of overproduction. Combined with local renewable generation, decentralized storage would also have the advantage of improving the robustness of the electricity network by allowing islanding of the area supplied by this resource. Also, a well-placed ESS increases the quality of the power supplied by providing better control of frequency and voltage and reduces the impact of its variability by adding value to the current supplied, especially if the electricity is delivered during peak periods [5,6].

The integration of renewable energies together with the ESS from a standalone microgrid. In general, integrating various renewable energies such as tidal, wind, and photovoltaic (PV) to provide a favorable impact on the ESS's maximum capacity is recommended. Usually, ESS is constituted by a combination of a battery and supercapacitors, which helps extend battery life-time and offers a fast system response to compensate the transients [7]. However, loads are necessary when all [energy sources and battery storage systems (BSSs)] are connected, thus the AC grid is used instead of supercapacitors [8]. A microgrid, is classified into DC, AC, or a combination of both types. Compared with AC microgrid, DC microgrid shows several benefits such as fewer parameters to control, facilitate integration, and simple structure. On the other hand, AC type needs more information like the synchronization of the frequency and reactive power, which makes the control design process a challenging task. Moreover, DC microgrids offers the possibility to work in different modes like AC microgrid, standalone, or integrated with the AC microgrid [9,10].

The autonomous DC microgrid can operate at optimal efficiency thanks to recent advancements in power electronics. However, because of the RESs stochastic nature, the smooth operation and continuous power transmission to the loads need a supplementary energy management unit (EMU). Numerous researches work on the energy management control dedicated to AC microgrids can be found in the literature, but given the important differences between the AC and DC microgrid dynamics, these control strategies cannot be adopted for DC microgrids. In fact, in the standard design of the DC microgrid, the load converters and the energy sources are parallelly connected where the energy is consumed or supplied through the DC-link. As a result, regulation of the DC-link voltage is required for the DC microgrid to operate efficiently and reliably. Several control strategies have appeared in the literature to address the issues of the DC-link voltage. In Ref. [11], a review of the recent trends and development in hybrid microgrid with energy resource planning and control. In Ref. [12], a combined fuzzy controller and voltage control are proposed to regulate the DC voltage. In Ref. [13], a fuzzy logic control strategy with reduced rules is investigated. In Ref. [14], a dual proportional-integral (PI) controller is adopted. However, the aforementioned control strategies are linear and can regulate the DC-link in a small operating interval. Thus to overcome this restriction, nonlinear controls have been investigated in the literature. In Ref. [15], an adaptive droop controller algorithm is proposed. Energy management-based optimal control is investigated in Ref. [16] for multiple ESS in a microgrid. In Ref. [17], robust $H\infty$ control strategy is developed. Robust sliding mode strategy is proposed in Ref. [18]. In Ref. [19], an adaptive backstepping control method is designed. A Lyapunov-based strategy in Ref. [20]. Feedback linearization control in Ref. [21]. A hybrid combined backstepping and sliding mode is investigated in Ref. [22]. However, the previous proposed nonlinear controls show limitations in performances in the case of droop control strategy and optimal energy management has given the multiple integrated, poor stability for the $H\infty$ method, chattering issues concerning the sliding mode. Furthermore, a large portion of these controls rely on fixed gains, which are extremely susceptible to parameter inaccuracies and external disturbances. Finally, the EMU is represented in the final section. In the same vein, the current study proposes a novel modified super twisting algorithm (MSTA) combined with a fuzzy logic method to overcome the issues with traditional integer controls in hybrid energy management. Fig. 4.1 shows the suggested new controller in conjunction with an EMU for a DC-microgrid with various stochastic sources and important DC loads. The proposed intelligent modified super twisting algorithm (IMSTA) control will be used as a low-level controller, when the EMU serves as high-level controller which generates appropriate references for the IMSTA and monitors the generated and consumed power.

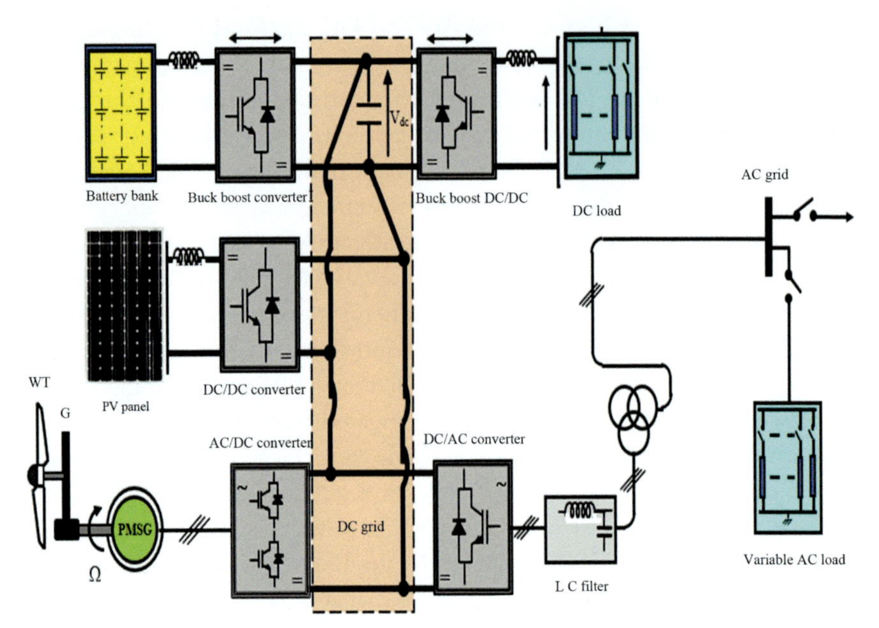

Figure 4.1 Studied hybrid system structure.

The major goals of this study are to use the suggested IMSTA to manage the source-side converters (SSCs) in order to extract the maximum power from the PV and wind sources. The second one is to improve the power quality supplied to the DC-microgrid by employing the EMU to regulate the DC-link voltage and reactive power to their respective references. The following is a summary of the work's novelty and contribution:

- For a DC-microgrid with many stochastic sources and important DC loads, a new MSTA controller paired with a fuzzy logic strategy is proposed.
- The fuzzy logic method is chosen as a fuzzy gain supervisor to adaptively adjust MSTA gains, considerably improving the proposed approach's robustness against various uncertainties and external disturbances.
- The essential characteristic of this approach is the extremely reduced number of the fixed gains used by the proposed strategy which avoids its sensitivity to parameter uncertainties, which highly improves the robustness property and global stability of the system.
- The system's overall stability is guaranteed and further proven thorough simulation results.

The current chapter is designed so that Section 4.2 contains a mathematical description of the hybrid energy system. The proposed hybrid controller strategy is designed in Section 4.3. The numerical results are provided and discussed in Section 4.4. Section 4.5 concludes with the key results and future work.

4.2 Mathematical description of the hybrid energy system

The studied hybrid energy system integrated smart DC-microgrid is illustrated by Fig. 4.1, where three main parts can be distinguished: the hybrid energy sources constituted by the wind energy, solar energy, and the BSSs connected to the DC-link through their respective converters. The second part represents the loads assumed to be a priority which in the case of a smart university may include laboratory experimentation benches, fans, and lighting. A maximum power point tracking algorithm is used on both the wind and solar (PV) conversion systems to force them to operate at maximum power. The EMU computes the total consumed and produced energy to select the adequate control modes.

4.2.1 Wind system model

The wind power mathematical model that the turbine can transform is given as [23]:

$$P_m = \frac{1}{2}\rho C_p(\beta, \lambda) A v_t^3 \tag{4.1}$$

$$T_m = \frac{P_m}{\omega_m} \tag{4.2}$$

$$C_p(\beta, \lambda) = 0.5\left(\frac{116}{\lambda_i} - 0.4\beta - 5\right)e^{-\left(\frac{21}{\lambda_i}\right)} \tag{4.3}$$

$$\lambda_i^{-1} = (\lambda + 0.08\beta)^{-1} - 0.035(1+\beta^3)^{-1} \tag{4.4}$$

$$\lambda = \frac{\omega_m R}{v_t} \tag{4.5}$$

where, v denotes the wind speed, β depicts the angle of the pitch, ω_t denotes the turbine speed, R represents the blades radius, C_p denotes the power coefficient, λ denotes the tip-speed ratio, ρ denotes the water density, and A represents the area of the blades. The wind conversion system is based on a permanent magnet synchronous generator which is expressed as Tamalouzt [23]:

$$v_{\mathrm{dq}} = R_{\mathrm{dq}} i_{\mathrm{dq}} + L_{\mathrm{dq}} \dot{i}_{\mathrm{dq}} + \psi_{\mathrm{dq}} p \omega_m \tag{4.6}$$

$$J\dot{\omega}_m = T_m - T_e - f_{fv}\omega_m \tag{4.7}$$

$$T_e = \frac{2}{3} p \psi_{\mathrm{dq}}^T i_{\mathrm{dq}} \tag{4.8}$$

where, $\mathrm{L_{dq}} = \begin{bmatrix} \mathrm{L_d} & 0 \\ 0 & \mathrm{L_q} \end{bmatrix}$ denotes the matrix of the inductances, T_e denotes the electromagnetic torque,$\mathrm{R_{dq}} = \begin{bmatrix} \mathrm{R_s} & 0 \\ 0 & \mathrm{R_s} \end{bmatrix}$ denotes the stator resistance matrix,$\mathrm{i_{dq}} = \begin{bmatrix} \mathrm{i_d} \\ \mathrm{i_q} \end{bmatrix}$ represents the vector of the stator current, $\mathrm{f_{fv}}$ represents the viscous friction coefficient, J is the moment of inertia, $\psi_{\mathrm{dq}} = \begin{bmatrix} \psi_{\mathrm{f}} \\ 0 \end{bmatrix}$ represents the flux linkages vector, and $\mathrm{v_{dq}} = \begin{bmatrix} \mathrm{v_d} \\ \mathrm{v_q} \end{bmatrix}$ represents voltage stator vector. To design the proposed control method, the model of the source-side converters (SCCs) needs to be expressed. Thus the model of the wind source converter (see Fig. 4.2) is given as Refs. [24,25]:

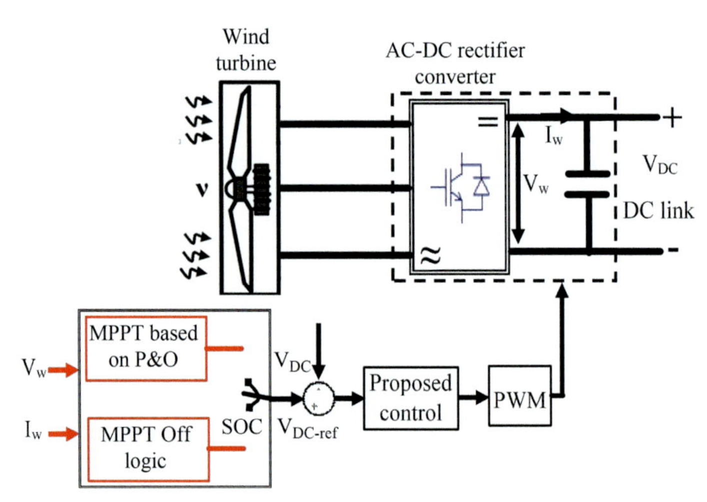

Figure 4.2 Wind energy system with controller.

$$\frac{dV_w}{\mathrm{dt}} = \frac{I_w}{C_w} - \frac{I_{\mathrm{Lw}}}{C_w} \tag{4.9}$$

$$\frac{V_w}{L_w} = \frac{dI_w}{\mathrm{dt}} + (1 - U_1)\frac{V_{\mathrm{dc}}}{L_w} - D_1 \tag{4.10}$$

$$\frac{dV_{\mathrm{dc}}}{\mathrm{dt}} = (1 - U_1)\frac{I_{\mathrm{Lw}}}{C_{\mathrm{dc}}} - \frac{I_{\mathrm{Ow}}}{C_{\mathrm{dc}}} + D_2 \tag{4.11}$$

where, I_w denotes the wind current rectified, L_w denotes the inductance, I_{Lw} denotes the current of the inductor, V_w denotes the voltage input rectified, U_1 denotes the control signal, V_{dc} denotes the link voltage, D_1 and D_2 denotes dynamics uncertainty in the energy stage parameters. The wind system can be run under maximum power point tracking (MPPT) for maximum power extraction or off-MPPT for power balance, depending on the status of the storage system, which will be detailed in the energy management section. The MPPT method is depicted in Fig. 4.3 as a flowchart.

The suggested EMU shifts the wind controller from MPPT mode to off-MPPT mode in order to lower generated power and maintain a

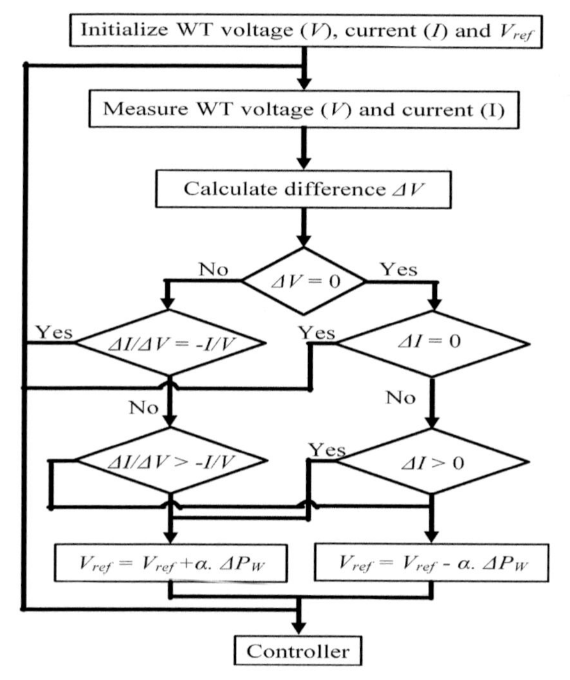

Figure 4.3 MPPT algorithm of the wind system.

balanced power in the standalone system when power generation exceeds storage capacity in the battery system. The voltage set-value is computed as follows in off-MPPT [21]:

$$V_{\text{ref}} = \frac{P_L - P_w}{I_w} \tag{4.12}$$

where, P_L denotes the power of the load and Pw denotes the generated wind power.

4.2.2 Solar power system model

The solar conversion system (SCS) is constituted by the PV panel connected to the DC-link through a DC-DC boost converter. The SCS mathematical model is given as below:

$$\frac{dV_{\text{pv}}}{\text{dt}} = \frac{I_{\text{pv}}}{C_{\text{pv}}} - \frac{I_{\text{Lpv}}}{C_{\text{pv}}} \tag{4.13}$$

$$\frac{V_{\text{pv}}}{L_{\text{pv}}} = \frac{\text{dI}_{\text{pv}}}{\text{dt}} + (1 - U_2)\frac{V_{\text{dc}}}{L_{\text{pv}}} - D_3 \tag{4.14}$$

$$\frac{dV_{\text{dc}}}{dt} = (1 - U_2)\frac{I_{\text{Lpv}}}{C_{\text{dc}}} - \frac{I_{\text{Opv}}}{C_{\text{dc}}} + D_4 \tag{4.15}$$

where, I_{pv} denotes the PV current, L_{pv} denotes the inductance, V_{pv} denotes the voltage of the PV panel, L_{Lpv} denotes the current of the inductor, U_2 denotes the control signal, D_3 and D_4 denotes dynamics uncertainty in the energy stage parameters as given by Fig. 4.3. Here also, according to the state of the storage system, for power balance, the PV conversion system can be run on MPPT or off-MPPT as shown in Figs. 4.4 and 4.5. Furthermore, if power generation exceeds battery storage capacity, the suggested energy management system shifts the PV controller from MPPT to off-MPPT mode to minimize generated power and maintain balanced power in the standalone system. In off-MPPT, the reference voltage is computed by Li et al. [21]:

$$V_{\text{ref}} = \frac{P_L - P_{\text{pv}}}{I_{\text{pv}}} \tag{4.16}$$

where, P_{pv} denotes the PV panel power.

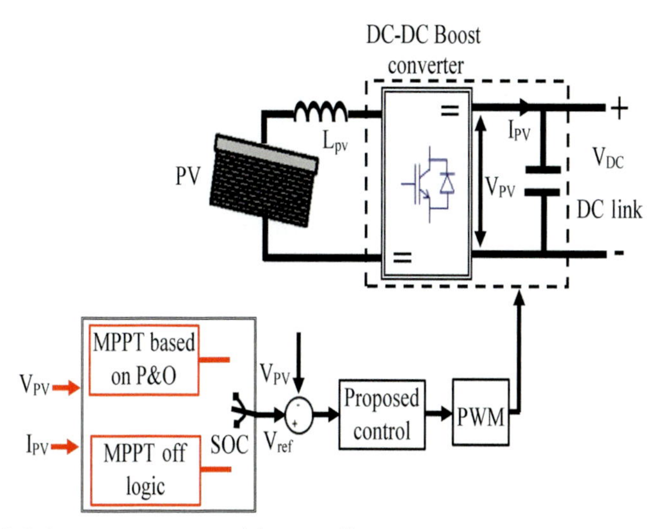

Figure 4.4 Solar energy system with controller.

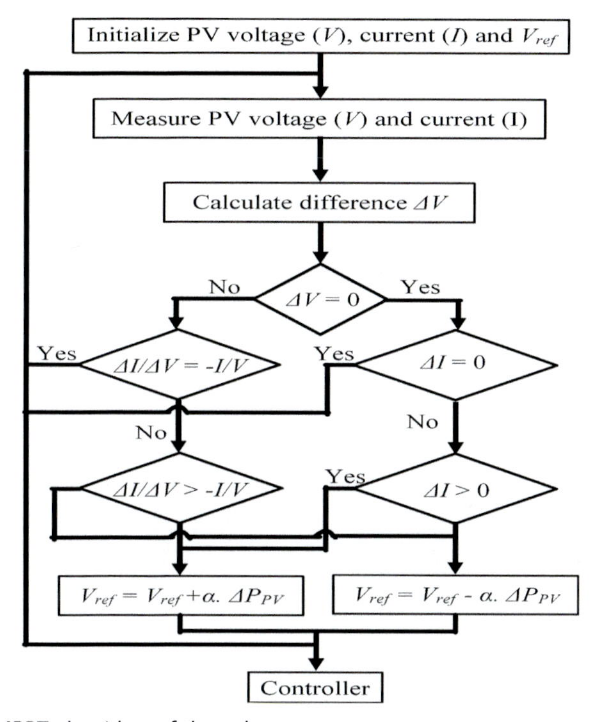

Figure 4.5 MPPT algorithm of the solar energy system.

4.2.3 Battery system model

A conventional battery is connected to the DC-link in this application via a bidirectional DC-DC back-boost converter connected to the micro-grid's DC-link (see Fig. 4.6). This converter's job is to keep the DC-link voltage constant regardless of power variations in the sources and load. The DC-link voltage is regulated at it references to compute the reference current of the battery and then design the voltage controller through the proposed strategy as shown in Fig. 4.6. The state of charge (SOC) of the battery is described by Singh et al. [24]:

The SOC, the amount of electricity stored during the charge, is an important parameter to be controlled. The battery SOC must detect by the proposed supervisory system to make decisions according to its status and the required power. In a battery, the ampere-hours stored during a time t corresponds to a nominal capacity Q and a charging current I_{bat}. The charge-discharge rate of a battery is determined by the available power, demand, and SOC. The SOC limitations are used to establish the battery's energy constraints as:

$$SOC = 100\left(1 + \frac{\int I_{bat}dt}{Q}\right) \tag{4.17}$$

$$SOC_{min} \leq SOC \leq SOC_{max} \tag{4.18}$$

where, SOC_{min} and SOC_{max} are the minimum and the maximum allowable states for the battery safety. The model of the BSS converter is given as:

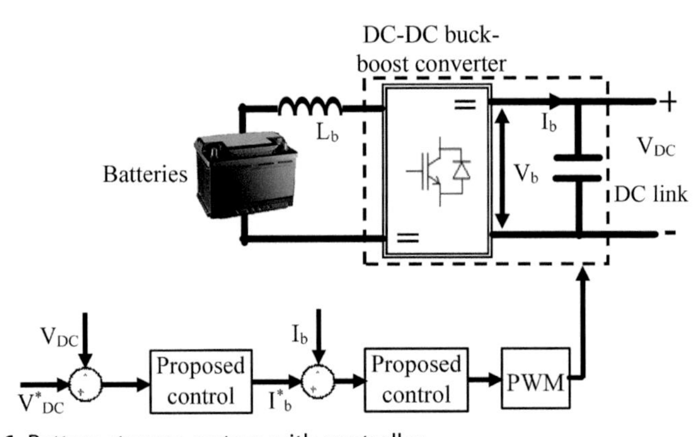

Figure 4.6 Battery storage system with controller.

$$\frac{V_b}{L_b} = \frac{dI_b}{dt} + U_3\frac{V_{\text{dc}}}{L_b} - D_5 \tag{4.19}$$

$$\frac{dV_{\text{dc}}}{dt} = U_3\frac{I_b}{C_{\text{dc}}} - \frac{I_{\text{Ob}}}{C_{\text{dc}}} + D_6 \tag{4.20}$$

where, I_b denotes the current of the battery, V_b denotes the voltage of the battery, U_3 denotes the controller signal, D_5 and D_6 denotes dynamics uncertainty in the energy stage parameters. The maximum power allowed during the charge/discharge of the battery system is fixed to 6525 W during the charge phase and to 10,440 W during the discharge phase, according to the parameters of the BSS given by Alharbi et al. [25].

4.2.4 AC grid model

Similar wind and AC grid converters are used which is a buck-to-buck converter (see Fig. 4.7). Then, the mathematical modeling of the AC grid converter system can be expressed as given below:

$$\frac{dV_g}{dt} = \frac{I_g}{C_g} - \frac{I_{\text{Lg}}}{C_g} \tag{4.21}$$

$$\frac{V_g}{L_g} = \frac{dI_g}{dt} + (1 - U_4)\frac{V_{\text{dc}}}{L_g} - D_7 \tag{4.22}$$

$$\frac{dV_{\text{dc}}}{dt} = (1 - U_4)\frac{I_{\text{Lg}}}{C_{\text{dc}}} - \frac{I_{\text{Og}}}{C_{\text{dc}}} + D_8 \tag{4.23}$$

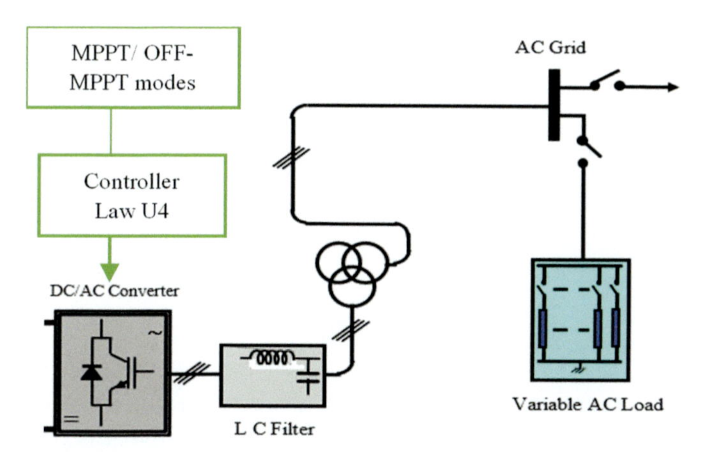

Figure 4.7 AC load system.

where, I_g denotes the rectified current of the grid, V_g denotes the voltage input rectified of the grid to the boost converter, U_4 denotes the controller signal, D_7 and D_8 denotes dynamics uncertainty in the energy stage parameters, L_{Lg} denotes the current of the inductor, and C_{dc} denotes the DC-link capacitor. From the model Eqs. (4.9)–(4.15) and model Eqs. (4.21)–(4.23), a generalized compact form can be deduced as given below:

$$\frac{dV_j}{\mathrm{dt}} = \frac{I_j}{C_j} - \frac{I_{\mathrm{Lj}}}{C_j} \tag{4.24}$$

$$\frac{V_j}{L_j} = \frac{dI_j}{\mathrm{dt}} + (1 - U_i)\frac{V_{\mathrm{dc}}}{L_j} - D_i \tag{4.25}$$

$$\frac{dV_{\mathrm{dc}}}{\mathrm{dt}} = (1 - U_i)\frac{I_{\mathrm{Lj}}}{C_{\mathrm{dc}}} - \frac{I_{\mathrm{Oj}}}{C_{\mathrm{dc}}} + D_{i+1} \tag{4.26}$$

where, the subscript j denotes the given sub-terms w, pv, b, and g of each converter. The subscript i denotes 1 in case of wind, 2 in case of PV, 3 in case of Battery, and 4 in case of AC grid.

Remark 1:. It should be remaindered that in the present work the MPPT algorithm is only used on SSCs. As a result, the grid energy supply is considered to be constant.

4.2.5 Load side converters model

In Fig. 4.1 it can be seen that a parallel DC-DC buck converter is used to connect the DC priority loads which their power loads are constant. These parallel converters are adopted to minimize the converters' stress and divide the load. The mathematical model of several parallel converters is given as below:

$$\frac{U_p V_{dc}}{L_p} = \frac{dI_{Lp}}{dt} + \frac{V_{Loadp}}{L_p} - D_{I_{Lp}} \tag{4.27}$$

$$\frac{dV_{Loadp}}{dt} = \frac{I_{Lp}}{C_p} - \frac{V_{Loadp}}{R_{Lp} C_c} + D_{V_{Loadp}} \tag{4.28}$$

where, U_p denotes the control law, L_{Lp} denotes the current of the inductor, V_{Loadp} denotes the load voltage, $D_{V_{\mathrm{Loadp}}}$ denotes dynamics uncertainty of the voltage, and $D_{I_{\mathrm{Lp}}}$ denotes dynamics uncertainty in the current.

4.3 Mathematical description of the hybrid energy system

The purpose of the proposed IMSTA is to compute the SCCs controller law shown in the generalized model Eqs. (4.24)–(4.26) represented by U_i and to compute the load side converters (LSCs) controller law shown in the generalized model Eqs. (4.27)–(4.28) represented by U_p as illustrated by Figs. 4.1–4.5.

The developed IMSTA must be designed in two steps: first, the controller laws are calculated by the MSTA and then, the fixed gains are adopted by the Fuzzy gain supervisor, which makes the proposed controller adaptive and robust against parameter uncertainties. To compute the controllers of the SSCs and LSCs, Singh et al. [24] proposes a PI control. Fixed gains, on the other hand, are notoriously difficult to compute when parameter uncertainties or variances exist [25]. As a result, the IMSTA is introduced to improve the resilience of the PI loops and to tackle their difficulties.

4.3.1 Source-side converters controllers design

To compute the SSCs controller law U_i, the following Lyapunov function as:

$$V_{j1} = 0.5e_j^2 \tag{4.29}$$

where, $e_j = C_j(V_j - V_j^*)$ is the voltage error and V_j^* denotes the desired voltage controller. From Eq. (4.24) and the derivative of e_j it yields:

$$e_j = I_j - I_{Lj} - C_j\dot{V}_j^* \tag{4.30}$$

To design the voltage controller, the desired current I_{Lj}^* is needed which is deduced from Eq. (4.30) as given below:

$$I_{Lj}^* = I_j - C_j\dot{V}_j^* + k_je_j \tag{4.31}$$

From Eq. (4.31) derivative of Eq. (4.29), it gives:

$$\dot{V}_{j1} = -k_je_j^2 \tag{4.32}$$

where, $k_j > 0$ is the gain matrix. The EMU generates the BSS desired current. Thus to design the controller law U_i, the MSTA is adopted as illustrated in Fig. 4.8. For more information about the MSTA and its

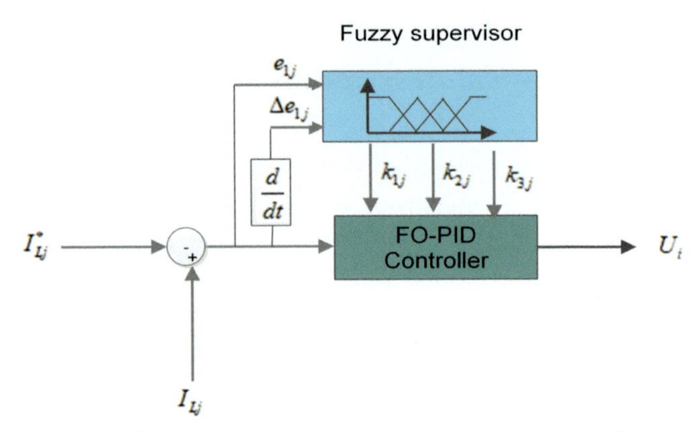

Figure 4.8 Source-side converters controller law computation with the modified super twisting algorithm.

benefits, the reader is directed to Ref. [26]. The expression of the controller law U_i is given as:

$$\begin{cases} U_i = k_{1j}\left|e_{1j}\right|^{0.5}\text{sign}(e_{1j}) + k_{2j}e_{1j} + u \\ u = k_{1j}\text{sign}(e_{1j}) + k_{2j}e_{1j} \end{cases} \tag{4.33}$$

where, k_{1j}, and k_{2j} denotes the gain matrix, and e_{1j} denotes the current error expressed as:

$$e_{1j} = (I_{\text{Lj}} - I^*_{\text{Lj}}) \tag{4.34}$$

where, the desired current I^*_{Lj} in Eq. (4.31) is computed using the MSTA as below:

$$\begin{cases} L^*_{Lj} = k_{1j}\left|e_{2j}\right|^{0.5}\text{sign}(e_{2j}) + k_{2j}e_{2j} + u \\ u = k_{1j}\text{sign}(e_{2j}) - k_{2j}e_{2j} \end{cases} \tag{4.35}$$

where, e_{2j} denotes the DC-link voltage error expressed as:

$$e_{2j} = (V_{\text{dc}} - V^*_{\text{dc}}) \tag{4.36}$$

Fixed gains k_{1j}, and k_{2j}, on the other hand, are extremely sensitive to parameter changes, as previously indicated. As a result, the fuzzy technique is chosen as the fuzzy supervisor is employed for gain adaption, solving the difficulty caused by imprecise parameters. The fuzzy inputs are chosen as the current error in the case of the controller law computation given by Eq. (4.33) and its derivative or the DC-link error in the case of the desired current given by Eq. (4.35) and its derivative. Triangular and trapezoidal types symmetrical and uniformly distributed are used to select the membership functions as given in Fig. 4.9. Lee and Takagi [27] and Yubazaki et al. [28] describe a strategy for splitting these functions. Their approach is built on the idea of numerous membership functions sharing the same parameter. The benefit of this strategy is that the number of membership function arguments is greatly decreased. The crisp outputs are generated using the center of gravity defuzzification approach, while the decision-making output is obtained using a Max-Min fuzzy inference. In Table 4.1, the linguistic variables corresponding to the fuzzy gain

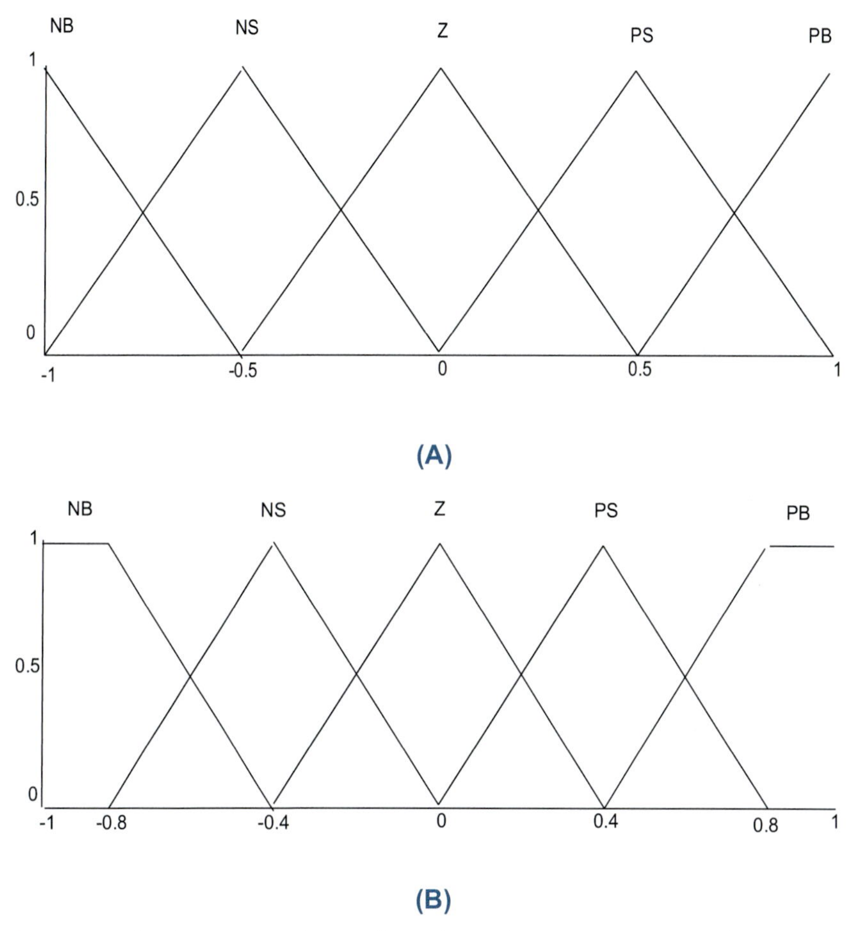

Figure 4.9 The fuzzy controller configuration. (A) Input membership function, (B) Output membership function.

Table 4.1 Fuzzy logic rules of the source-side converters.

$\frac{\Delta e_j}{e_j}$	NB	NS	Z	PS	PB
NB	NB	NB	NS	NS	Z
NS	NB	NB	NS	Z	PS
Z	NS	NS	Z	PS	PS
PS	NS	Z	PS	PB	PB
PB	Z	PS	PS	PB	PB

scheduling inputs-outputs are labeled as follows: Positive Big (PB), Negative Big (NB), Positive Small (PS), Negative Small (NS), and Zero (Z) [27].

Remark 2:. fixed gains are difficult to calculate when the system is exposed to parameter changes, hence the fuzzy logic controller was chosen to compute the gains of the super twisting method. Also, the stability and robustness proof of the super twisting algorithm are clearly demonstrated in Refs. [26] and [28], thus they are not considered in the present work.

4.3.2 Load side converters controller design

The expression of the controller law U_p is given as (see Fig. 4.10):

$$\begin{cases} U_p = k_{1p}\left|e_{1p}\right|^{0.5}\text{sign}(e_{1p}) + k_{2p}e_{1p} + u \\ u = k_{1p}\text{sign}(e_{1p}) + k_{2p}e_{1p} \end{cases} \tag{4.37}$$

where, k_{1p}, k_{2p}, and k_{3p} denotes the gain matrix, and e_{1p} denotes the current error expressed as:

$$e_{1p} = (I_{\text{Lp}} - I_{\text{Lp}}^*) \tag{4.38}$$

where, the desired current I_{Lp}^* in Eq. (4.38) is computed using the MSTA as below:

$$\begin{cases} L_{\text{Lp}}^* = k_{1p}\left|e_{2p}\right|^{0.5}\text{sign}(e_{2p}) + k_{2p}e_{2p} + u \\ u = k_{1p}\text{sign}(e_{2p}) - k_{2p}e_{2p} \end{cases} \tag{4.39}$$

where, e_{2p} denotes the load voltage error expressed as:

$$e_{2p} = (V_{\text{Loadp}} - V_{\text{Loadp}}^*) \tag{4.40}$$

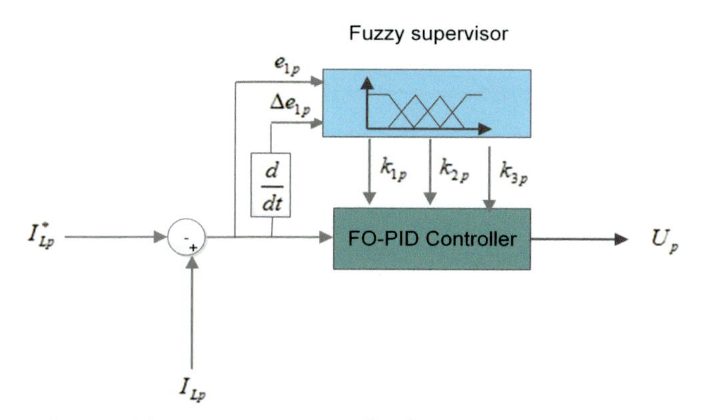

Figure 4.10 Source-side converters controller law computation with the modified super twisting algorithm.

4.3.3 Energy management unit

The EMU aim is to coordinate and control all the operations in the microgrid system. From Figs. 4.2–4.7, it can be seen that the EMU described by the MPPT Mode/of-MPPT mode algorithm is used to generate the references of the SSCs and LSCs controller law. The EMU generates the references based on the measured input power available and the consumed for both the SSCs and LSCs. The renewable sources are prioritized as mentioned previously on the loads.

The BSS works in charge/discharge mode and regulates the DC-link voltage at its reference value. the power in the microgrid is balanced under different power generation forms oh the renewable sources and the load demand condition. When the SSCs generate abundant power, the supply power is used to charge the BSS. In case the power generated by the SSCs is not enough, the power in the AC grid is used to supply the loads as shown in Fig. 4.1. The mathematical model of the power balance is given as Refs. [15,24]:

$$P_W + P_{\text{pv}} + P_g = P_{\text{Load}} + P_{\text{Battery}} \tag{4.41}$$

According to Fig. 4.11, four modes of the EMU can be distinguished and each mode depends on two conditions: the battery state and the generated power. When the generated power from the renewable energies is more than the load demand, the additional power is transferred to charge the battery to its SOC_{max}, at this limit the MPPT is switched to off-mode. In case the generated power cannot meet the load demand, the required power is supplied by the BSS until SOC_{min}, and in case the

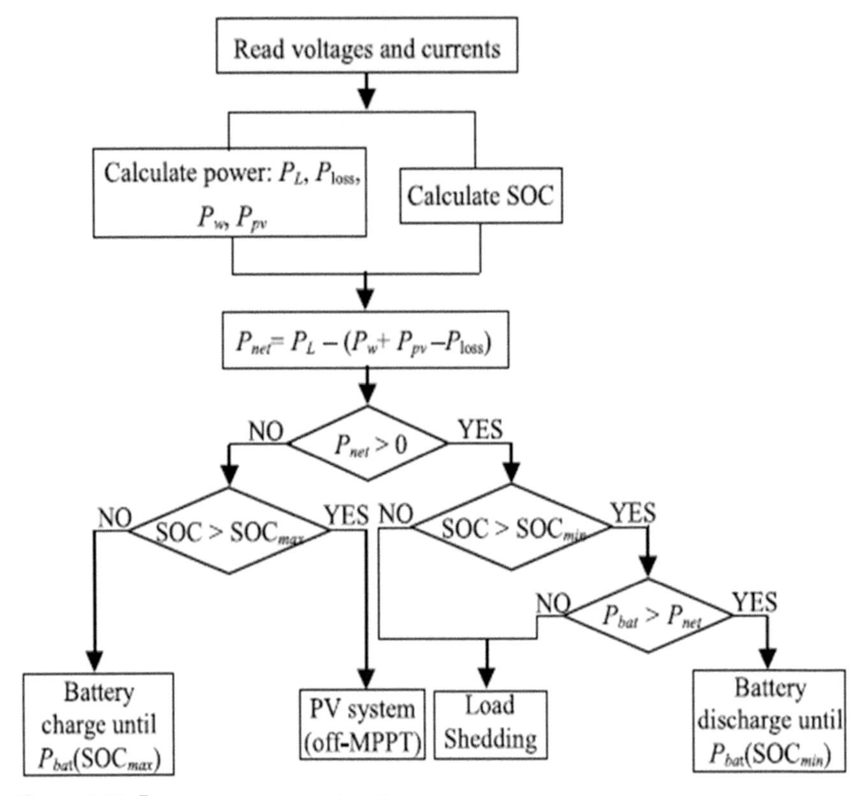

Figure 4.11 Energy management system.

power generated by the SSCs is not enough, the power in AC grid is used to supply the loads. Thus the generalized energy management controller structure is illustrated in Fig. 4.12.

4.4 Numerical results

The current chapter proposed a combined hybrid energy system integrated smart DC-microgrid, which is depicted in Fig. 4.1, with three major components: the hybrid energy sources comprised of wind energy, solar energy, and the BSS connected to the DC-link via their respective converters. In the case of a smart university, the second part represents the loads assumed to be a priority, which may include laboratory experimentation benches, fans, and lighting. On both wind and solar (PV) conversion systems, a maximum power point tracking algorithm is used to force them to operate at maximum power.

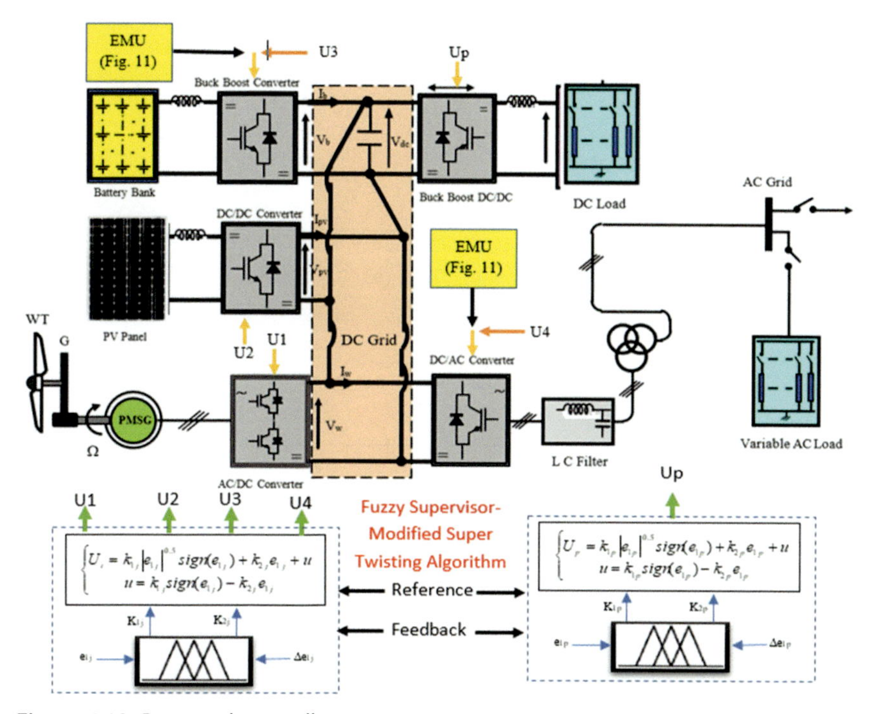

Figure 4.12 Proposed controller structure.

The proposed system's simulation results are generated using Matlab/Simulink, and the parameters used can be found in Ref. [25]. The DC-link reference voltage is set to 240 V. The simulation test is centered on the performance of the EMUs depicted in Fig. 4.11. When the BSS's SOC is initially at 80%, an 8000-W DC load is connected to the DC-link via two load-side converters. Fig. 4.13 depicts the wind profile between 8 and 12 m/s. Fig. 4.14 depicts the generated wind power, which varies (5000–10,000 W) with wind speed. Fig. 4.15 depicts the generated PV power, which is constant at 3000 W under a radiance of 600 W/m^2 and a temperature of 25°C. Fig. 4.16, depict the generated power P_{dg} from both PV and wind sources. From the present responses, one can see that the power P_{dg} increases between 7000 and 13,000 W.

Figs. 4.17 and 4.18 show the battery power and its SOC. From the presented results, the battery supplies the microgrid with about 2300 W in the time intervals [0–3.5]s, when SOC > 20%, while in the time intervals [3.5–10]s P_{dg} is more than the load power, Thus the battery is charged with about 4500 W from the microgrid. Fig. 4.19, shows the DC-link

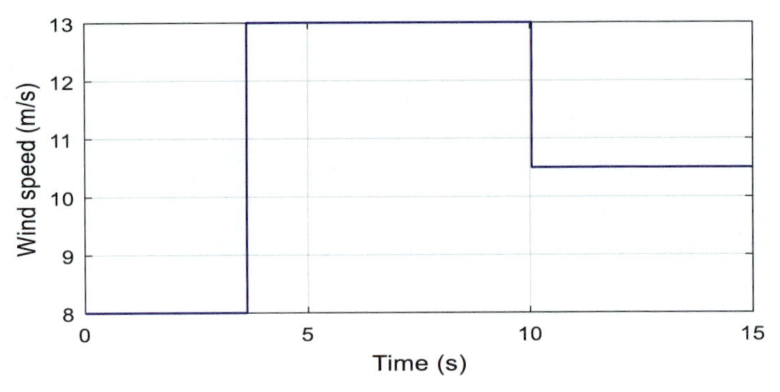

Figure 4.13 Wind speed.

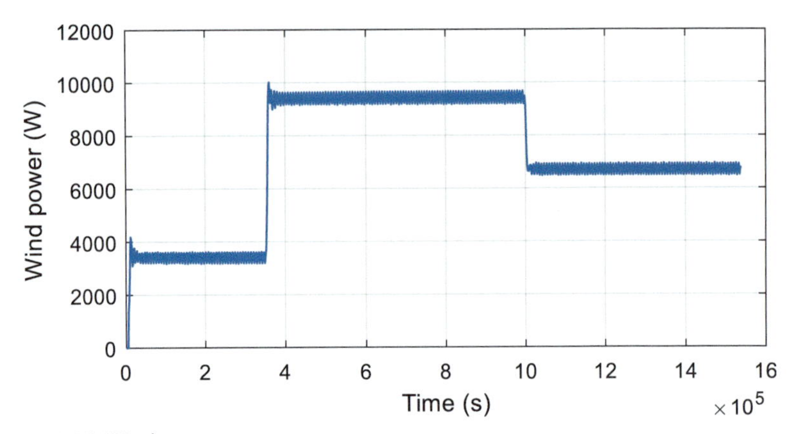

Figure 4.14 Wind power.

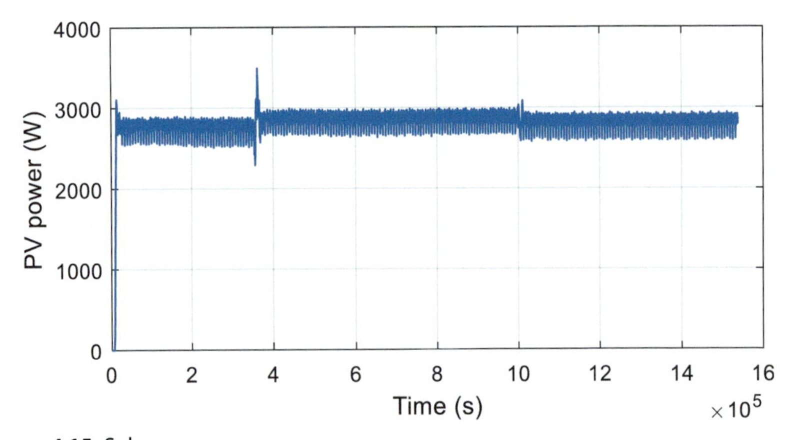

Figure 4.15 Solar power.

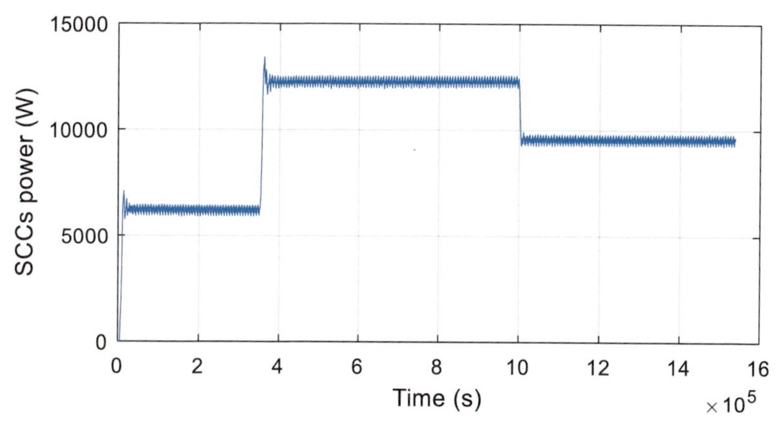

Figure 4.16 Source-side converters power.

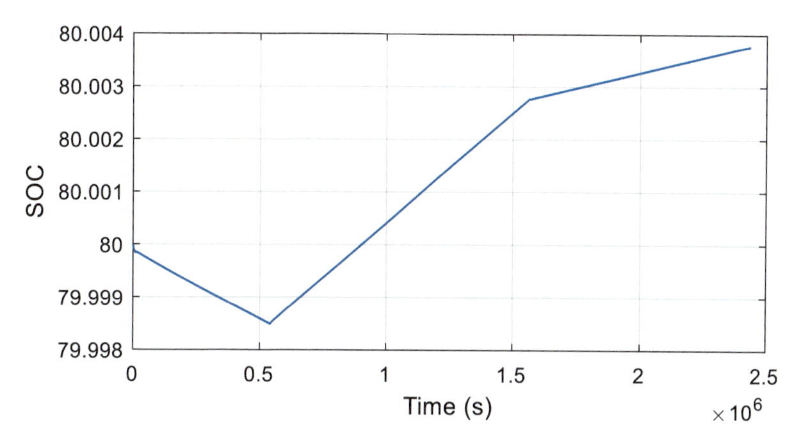

Figure 4.17 Battery storage system power.

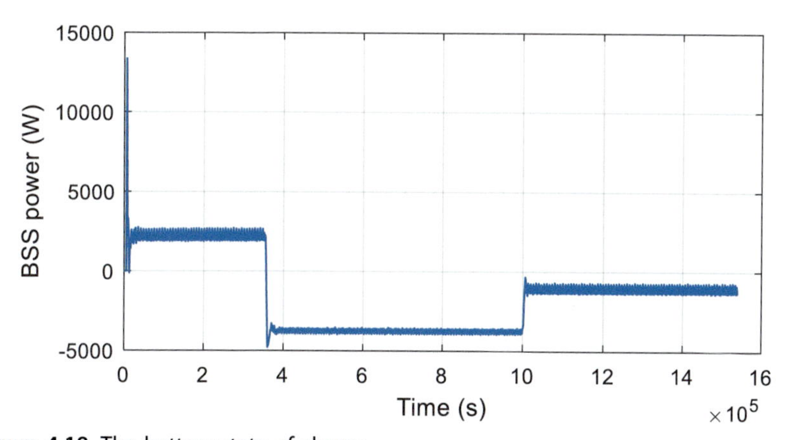

Figure 4.18 The battery state of charge.

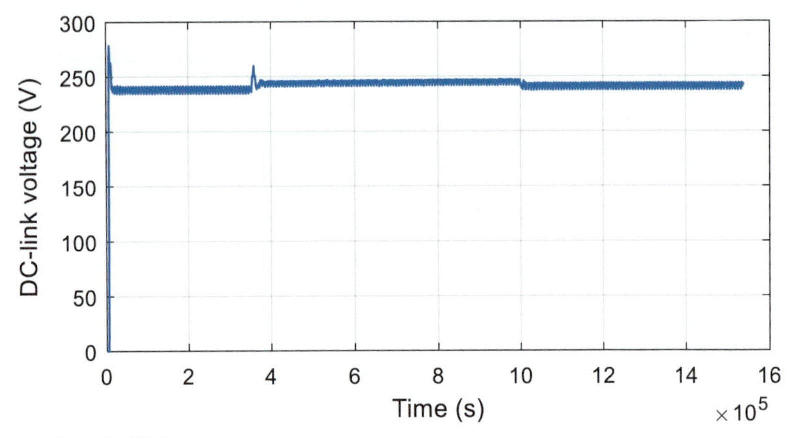

Figure 4.19 DC-link voltage.

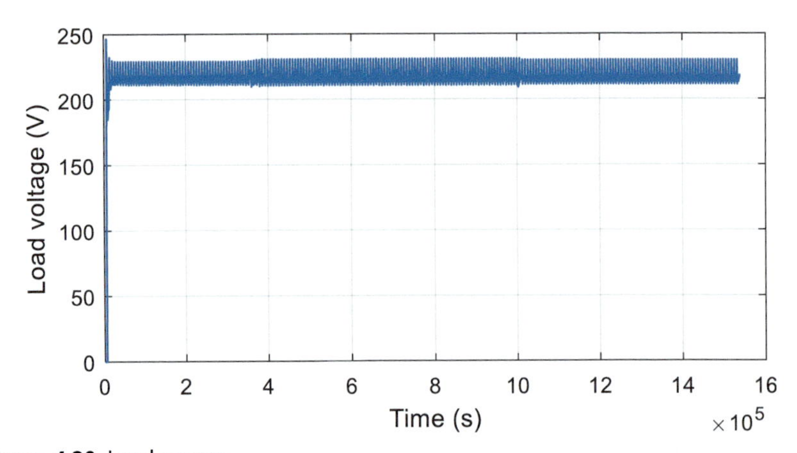

Figure 4.20 Load power.

voltage of both the SSCs and LSCs for the PI and proposed IMSTA, where it can be seen that both regulate the DC-link at its reference value. However, the proposed IMSTA shows superior performances in terms of the steady-state error and the convergence criterion. Fig. 4.20 shows that the proposed energy management controller transmits a constant power to the loads about 8300 W. Fig. 4.21 clearly indicates that the proposed IMSTA regulates the output voltage at its references (220 V) (Table 4.2).

A comparative analysis with previous works has been performed in the present section to highlight the advantages of the proposed IMSTA. The comparative analysis is shown by Table 4.3. Extensive comparative

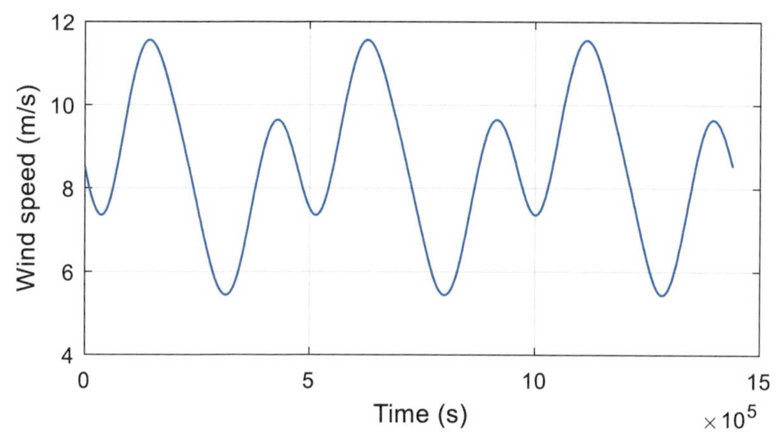

Figure 4.21 Load voltage.

Table 4.2 Fuzzy logic rules of the load side converters.

$\frac{\Delta e_p}{e_p}$	**NB**	**NS**	**Z**	**PS**	**PB**
NB	NB	NB	NS	NS	Z
NS	NB	NB	NS	Z	PS
Z	NS	NS	Z	PS	PS
PS	NS	Z	PS	PB	PB
PB	Z	PS	PS	PB	PB

analysis with Ref. [25], fractional order proportional-integral-derivative (FO-PID) and PID is demonstrated in Table 4.4, where it can be seen that the proposed strategy generates more power and shows high performance over the proposed control strategies. From the present comparative analysis, the proposed controller produces +3.15% wind power, +50% PV power, +2.5% load power over the super twisting fractional-order and more when compared to the PID control. In summary, the proposed controller has well managed the hybrid energy, and well achieved the objectives of the present work, and shows higher performances when compared to the other methods.

To test the robustness of the proposed energy management strategy, a random variation of the wind speed and solar radiance is used as shown by Figs. 4.22 and 4.23, respectively. Fig. 4.24 shows the wind power generated under random wind profile. From the presented results, it can be seen that the wind system works at MPPT. Same for the PV power which is maintained at the MPPT (see Fig. 4.25). Fig. 4.26 shows the

Table 4.3 Comparative analysis of the proposed strategy with recent references.

References	Microgrid elements	Method	Main contribution	The novelty of the proposed strategy
Kakigano et al. [12]	Wind-PV + BSS + DC loads (houses)	Distribution voltage control	FLC + gain scheduling	-A new adaptive and intelligent controller is proposed. The controller is used to control both SSCs and LSCs contrary. -The number of the fixed gains used by the proposed strategy is the extremely reduced (Zero fixed gains) as all the gains are computed by the fuzzy supervisor which avoid its sensitivity to parameter uncertainties, and thus highly improve the robustness property and global stability of the system. -Increases the power produced.
Aviles et al. [13]	PV-wind-BSS-residence	Energy management	Two low-complexity FLC	
Kumar et al. [14]	PV-wind-BSS-SOFC-loads	Coordinated control	Two feed-back control loops and feed-forward control loop	
Shaw et al. [15]	PV-BBSS-loads	Energy management	Adaptive droop control	
Kapoor et al. [19]	PV-wind generator-BSS-loads	Energy management	Adaptive backstepping	
Alharbi et al. [25]	PV-wind-BSS-load	Energy management + SSCs control	Super twisting fractional order	
Proposed strategy	Wind-PV-BSS-loads	Energy management + SSCs Control	Fuzzy supervisory-MSTA	

Table 4.4 Results comparison of the proposed strategy with that of Ref. [25].

Controller	Proposed IMSTA	Super twisting factional order [25]	FO-PID [26]	PID	
Wind power (W)	9800 (+3.15%)	9500	9800	9400	
PV power (W)	3000 (+50%)	2000	3000	1900	
SSCs power (W)	13,000 (+4%)	12,500	13,000	12,300	
BSS power stored (W)	2500 (+13.64%)	2200	2500	2100	
BSS power supplied (W)	4500 (+12.5%)	4000	4500	4000	
Load power (W)	8300 (+2.5%)	8100	8300	8000	
Complexity	Low	High	Low	Very Low	
Robustness	High (zero fixed gains)	Poor (more than seven fixed gains)	Low (more than five fixed gains)	Very poor (more than 10 fixed gains)	
Performance	Very high	High	High	Low	
PB	Z	PS	PS	PB	PB

power generated from both PV and wind sources. From the given results, it can be seen that the power generated is maintained between 5000 and 13,000 W which is the same as in the first test (see Fig. 4.16). Fig. 4.27 shows the BSS power under the random variations which is varying between 5000 and −5000 W. Fig. 4.28 shows the Battery SOC under random variations during the discharge mode. Fig. 4.29 depicts that the proposed energy management control transmits a constant power to the load (about 8300 W), here also it's the same as in the first case (see Fig. 4.20). Fig. 4.30 shows the DC-link voltage response. From the presented response, it can be seen that the proposed strategy well regulates the DC voltage at its references.

Thus the proposed energy management strategy well validates the objectives even under random variations. Fig. 4.31, shows the load voltage under random variations which is perfectly fixed at its reference value even under these changes.

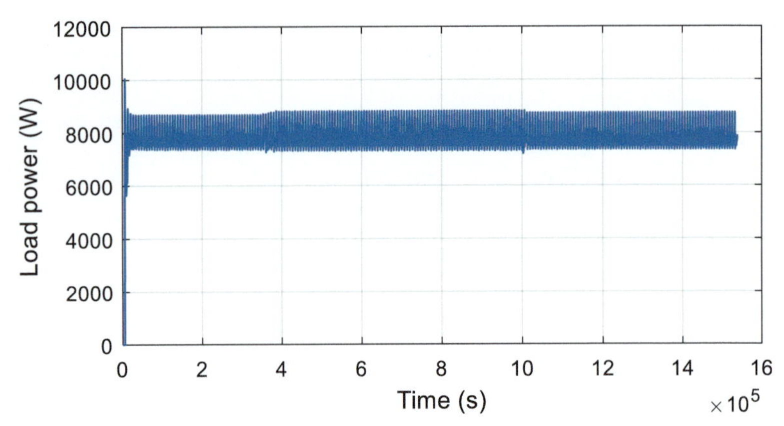

Figure 4.22 Random wind speed.

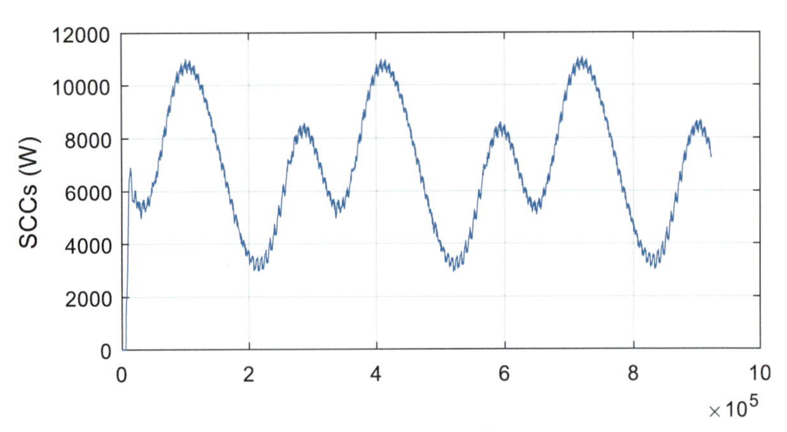

Figure 4.23 Wind power under random wind speed.

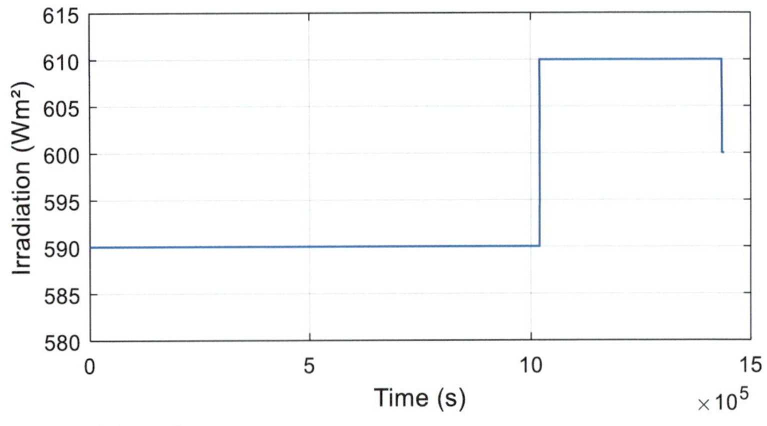

Figure 4.24 Solar radiance.

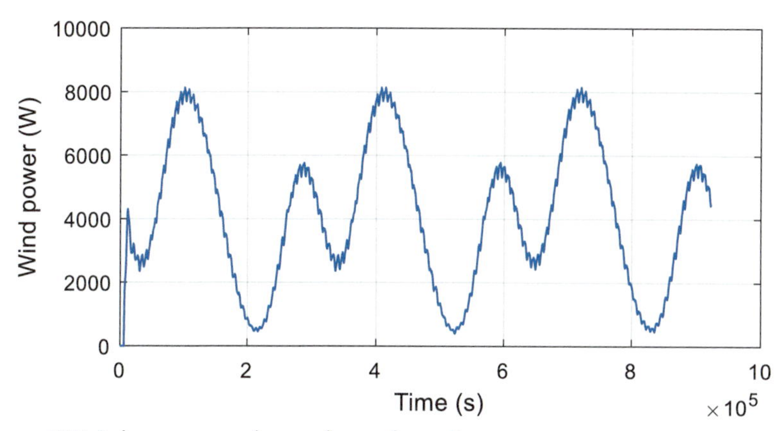

Figure 4.25 Solar power under random solar radiance.

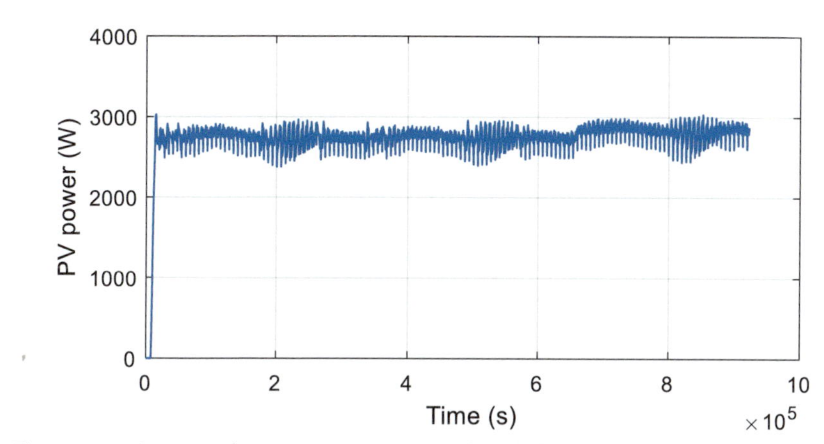

Figure 4.26 Source-side converters power under random variations.

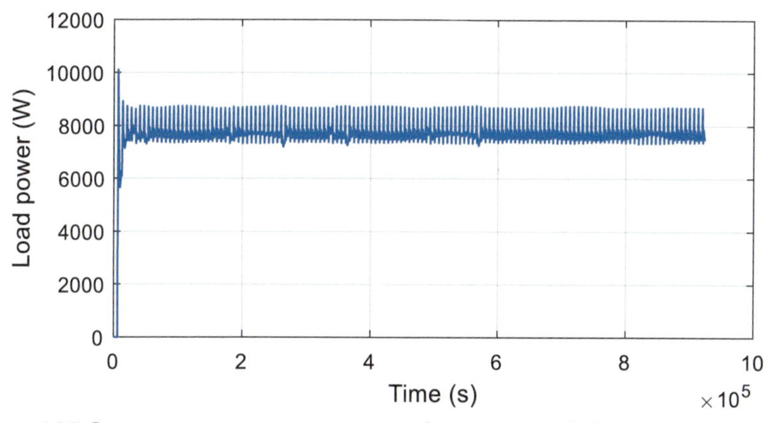

Figure 4.27 Battery storage system power under random variations.

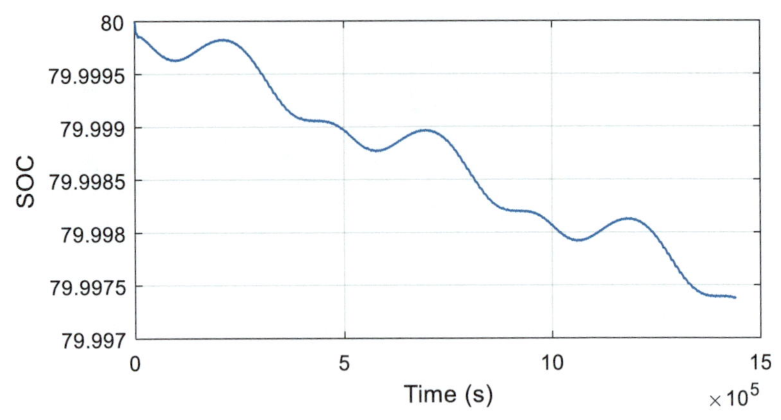

Figure 4.28 The battery state of charge under random variations.

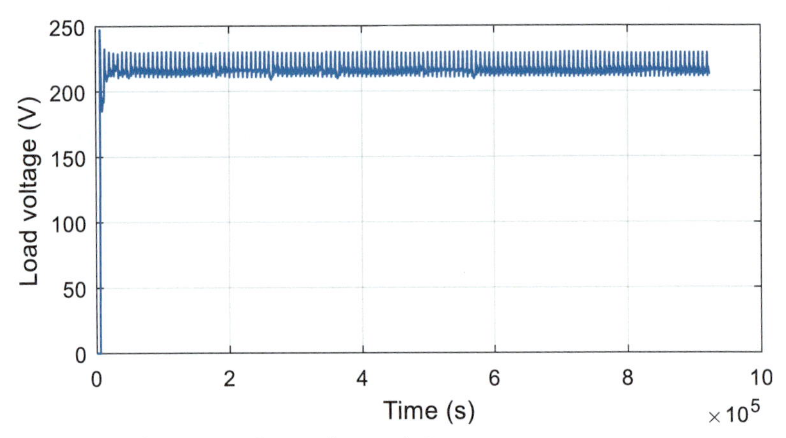

Figure 4.29 Load power under random variations.

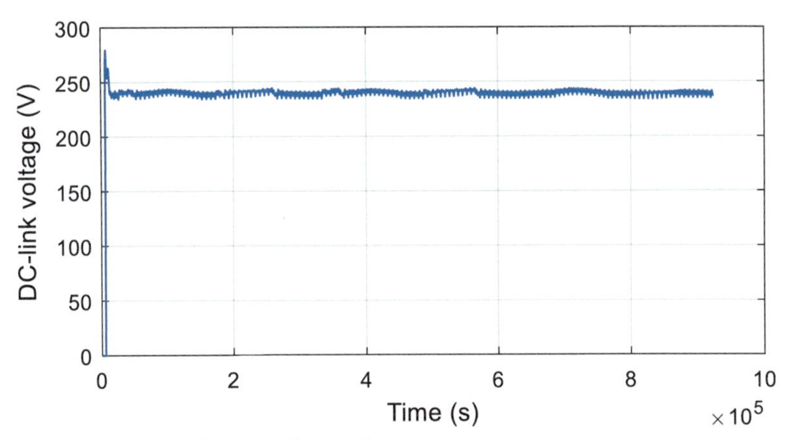

Figure 4.30 DC-link voltage under random variations.

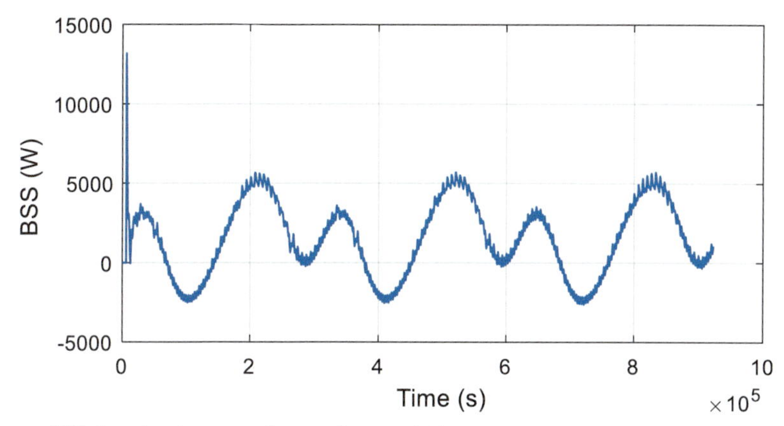

Figure 4.31 Load voltage under random variations.

4.5 Conclusion

In this chapter, a novel IMSTA is proposed for the Energy management of hybrid energy sources contacted to a smart grid through a DC-link voltage. The hybrid energy sources integrated to the DC-microgrid are constituted by a battery bank, wind energy, and PV energy source. The SCCs are controlled by the new IMSTA to extract the maximum power from the RESs (wind and PV) and improve the power quality supplied to the DC-microgrid. To make the microgrid as cost-effective, the (Wind and PV) energy sources are prioritized. The proposed controller ensures smooth output power and service continuity. Simulation results of the proposed control schema under Matlab/Simulink are presented and compared with the other nonlinear controls. Extensive comparative analysis with super twisting fractional order control, FO-PID and PID is demonstrated in Table 4.4, where it can be seen that the proposed strategy generates more power and shows high performance over the proposed control strategies. From the present comparative analysis, the proposed controller produces +3.15% wind power, +50% PV power, +2.5% load power over the super twisting fractional-order and more when compared to the PID control.

References

[1] Y. Allahvirdizadeh, M.P. Moghaddam, H. Shayanfar, A survey on cloud computing in energy management of the smart grids, Int. Trans. Electr. Energy Syst. 29 (10) (2019) e12094. Available from: https://doi.org/10.1002/2050-7038.12094.

[2] C. Byer, A. Botterud, Additional capacity value from synergy of variable renewable energy and energy storage, IEEE Trans. Sustain. Energy 11 (2) (2020) 1106–1109.
[3] M. Rizwan, L. Hong, W. Muhammad, S.W. Azeem, Y. Li, Hybrid Harris Hawks optimizer for integration of renewable energy sources considering stochastic behavior of energy sources, Int. Trans. Electr. Energy Syst. 31 (2) (2021) e12694. Available from: https://doi.org/10.1002/2050-7038.12694.
[4] Y. Sun, Z. Zhao, M. Yang, D. Jia, W. Pei, B. Xu, Overview of energy storage in renewable energy power fluctuation mitigation, CSEE J. Power Energy Syst. 6 (1) (2020) 160–173.
[5] T. Salameh, M.A. Abdelkareem, A.G. Olabi, E.T. Sayed, M. Al-Chaderchi, H. Rezk, Integrated standalone hybrid solar PV, fuel cell and diesel generator power system for battery or supercapacitor storage systems in Khorfakkan, United Arab Emirates, Int. J. Hydrog. Energy 46 (8) (2021) 6014–6027.
[6] M. Çolak, I. Kaya, Multi-criteria evaluation of energy storage technologies based on hesitant fuzzy information: a case study for Turkey, J. Energy Storage 28 (2021) 101211.
[7] M.A. Hannan, M.M. Hoque, A. Mohamed, A. Ayob, Review of energy storage systems for electric vehicle applications: issues and challenges, Renew. Sustain. Energy Rev. 69 (2017) 771–789.
[8] R. Amirante, E. Cassone, E. Distaso, P. Tamburrano, Overview of recent developments in energy storage: mechanical, electrochemical and hydrogen technologies, Energy Convers. Manag. 132 (2017) 372–387.
[9] T. Ma, H. Yang, L. Lu, Development of hybrid battery-supercapacitor energy storage for remote area renewable energy systems, Appl. Energy 153 (2017) 56–62.
[10] X. Wang, D. Yu, S. Blond, Z. Zhao, P. Wilson, A novel controller of a battery-supercapacitor hybrid energy storage system for domestic applications, Energy Buildings 141 (2017) 167–174.
[11] A.K. Barik, S. Jaiswal, D.C. Das, Recent trends and development in hybrid microgrid: a review on energy resource planning and control, Int. J. Sustain. Energy (2021). Available from: https://doi.org/10.1080/14786451.2021.1910698.
[12] H. Kakigano, Y. Miura, T. Ise, Distribution voltage control for DC microgrids using fuzzy control and gain-scheduling technique, IEEE Trans. Power Electron. 28 (5) (2013) 2246–2258.
[13] D.A. Aviles, J. Pascual, F. Guinjoan, G.G. Gutierrez, R.G. Orguera, J.L. Proano, et al., An energy management system design using fuzzy logic control: smoothing the grid power profile of a residential electro-thermal microgrid, IEEE Access 9 (2021) 25172–25188.
[14] M. Kumar, S.C. Srivastava, S.N. Singh, Control strategies of a DC microgrid for grid connected and islanded operations, IEEE Trans. Smart Grid 6 (4) (2015) 1588–1601.
[15] R.N. Shaw, P. Walde, A. Ghosh, IOT based MPPT for performance improvement of solar PV arrays operating under partial shade dispersion, in: 2020 IEEE 9th Power India International Conference (PIICON) held at Deenbandhu Chhotu Ram University of Science and Technology, SONEPAT, India on February 28–March 1, 2020, pp. 1–4, doi: 10.1109/PIICON49524.2020.9112952.
[16] Y. Xu, X. Shen, Optimal control based energy management of multiple energy storage systems in a microgrid, IEEE Access 6 (2018) 32925–32934.
[17] A.-R.-I. Mohamed, H.H. Zeineldin, M.M.A. Salama, R. Seethapathy, Seamless formation and robust control of distributed generation microgrids via direct voltage control and optimizeddynamic power sharing, IEEE Trans. Power Electron. 27 (3) (2012) 1283–1294.

[18] B.A.M. Trevino, A. El Aroudi, E.V. Idiarte, A.C. Pastor, R.M. Salamero, Sliding-mode control of a boost converter under constant power loading conditions, IET Power Electr. 12 (3) (2019) 521–529.
[19] G. Kapoor, P. Walde, R.N. Shaw, A. Ghosh, HWT-DCDI-based approach for fault identification in six-phase power transmission network, in: S. Mekhilef, M. Favorskaya, R.K. Pandey, R.N. Shaw (Eds.), Innovations in Electrical and Electronic Engineering. Lecture Notes in Electrical Engineering, vol. 756, Springer, Singapore, 2021. Available from: https://doi.org/10.1007/978-981-16-0749-3_29.
[20] Y. Belkhier, A. Achour, R.N. Shaw, A. Ghosh, Performance improvement for PMSG tidal power conversion system with fuzzy gain supervisor passivity-based current control, in: S. Mekhilef, M. Favorskaya, R.K. Pandey, R.N. Shaw (Eds.), Innovations in Electrical and Electronic Engineering. Lecture Notes in Electrical Engineering, vol. 756, Springer, Singapore, 2021. Available from: https://doi.org/10.1007/978-981-16-0749-3_6.
[21] X. Li, X. Zhang, W. Jiang, J. Wang, P. Wang, X. Wu, A novel assorted nonlinear stabilizer for DC–DC multilevel boost converter with constant power load in DC microgrid, IEEE Trans. Power Electron 35 (10) (2020) 11181–11192.
[22] G. Kapoor, V.K. Mishra, R.N. Shaw, A. Ghosh, Fault detection in power transmission system using reverse biorthogonal wavelet, in: S. Mekhilef, M. Favorskaya, R.K. Pandey, R.N. Shaw (Eds.), Innovations in Electrical and Electronic Engineering. Lecture Notes in Electrical Engineering, vol. 756, Springer, Singapore, 2021. Available from: https://doi.org/10.1007/978-981-16-0749-3_28.
[23] S. Tamalouzt, Performances of direct reactive power control technique applied to three level-inverter under random behavior of wind speed, Rev. Roum. Sci. Techn. – Electrotech. Energy 64 (1) (2019) 33–38.
[24] P. Singh, S. Bhardwaj, S. Dixit, R.N. Shaw, A. Ghosh, Development of prediction models to determine compressive strength and workability of sustainable concrete with ANN, in: S. Mekhilef, M. Favorskaya, R.K. Pandey, R.N. Shaw (Eds.), Innovations in Electrical and Electronic Engineering. Lecture Notes in Electrical Engineering, vol. 756, Springer, Singapore, 2021. Available from: https://doi.org/10.1007/978-981-16-0749-3_59.
[25] Y.M. Alharbi, A.A. Al Alahmadi, N. Ullah, H. Abeida, M.S. Soliman, Y.S.H. Khraisat, Super twisting fractional order energy management control for a Smart University System integrated DC micro-grid, IEEE Access 8 (2020) 128692–128704.
[26] Y. Belkhier, A. Achour, R.N. Shaw, W. Sahraoui, A. Ghosh, Adaptive linear feedback energy-based backstepping and PID control strategy for PMSG driven by a grid-connected wind turbine, in: S. Mekhilef, M. Favorskaya, R.K. Pandey, R.N. Shaw (Eds.), Innovations in Electrical and Electronic Engineering. Lecture Notes in Electrical Engineering, vol. 756, Springer, Singapore, 2021. Available from: https://doi.org/10.1007/978-981-16-0749-3_13.
[27] M.A. Lee, H. Takagi, Dynamic control of genetic algorithms using fuzzy logic techniques, in: Proceedings of International Conference on Genetic Algorithms, San Mateo, CA, 1993, pp. 76–83.
[28] Yubazaki N., Otami M., Ashid T., Hirota K., Dynamic fuzzy control method and its application to positioning of induction motor, Proceedings of 1995 IEEE International Conference on Fuzzy Systems., IEEE. 1995, pp. 1095-1102, vol. 3, doi: 10.1109/FUZZY.1995.4.

Further reading

A.K. Barik, D.C. Das, A. Latif, S.M.S. Hussain, T.S. Ustun, Optimal voltage–frequency regulation in distributed sustainable energy-based hybrid microgrids with integrated resource planning, Energies 14 (2021) 2735. Available from: https://doi.org/10.3390/en14102735.

A. Kadri, H. Marzougui, A. Aouiti, F. Bacha, Energy management and control strategy for a DFIG wind turbine/fuel cell hybrid system with super capacitor storage system, Energy 192 (2020) 116518.

CHAPTER FIVE

IoT in renewable energy generation for conservation of energy using artificial intelligence

Anand Singh Rajawat[1], Kanishk Barhanpurkar[2], Rabindra Nath Shaw[3] and Ankush Ghosh[4]

[1]Department of Computer Science Engineering, Shri Vaishnav Vidyapeeth Vishwavidyalaya, Indore, India
[2]Department of Computer Science, State University of New York, Binghamton, NY, United States
[3]Department of International Relations, Bharath Institute of Higher Education and Research (Deemed to be University), Chennai, India
[4]School of Engineering and Applied Sciences, The Neotia University, Sarisha, India

5.1 Introduction

An smart, dependable, and efficient power supply contributes to the smooth operation of smart cities. The use of internet of things (IoT) [1] in renewable energy [2] production will help smart cities meet their energy needs more efficiently. Sensors attached to generation, transmission, and distribution equipment are used in IoT applications in renewable energy [3] production. These gadgets let businesses remotely monitor and control the operation of their equipment in real-time. This reduces operational costs while also reducing our reliance on already scarce fossil resources. The IoT contains billions of interconnected things that can transfer information over a network without human interaction. The utilization and efficiency of renewable energy are aided by IoT infrastructure technology. Using IoT, users can combine smart city IoT device [4], and windows into a single system. Users can manage the operation of their electric equipment [5] via a PC or mobile app. Customers can also install clever gadgets in their homes or offices that measure energy [6] usage by specific equipment. You can utilize this data to identify waste and, in particular, power-hungry devices to save energy. Other IoT devices [7], such as thermostats, can automatically adjust their operation based on the climate to

Applications of AI and IOT in Renewable Energy.
DOI: https://doi.org/10.1016/B978-0-323-91699-8.00005-X

save energy. Residential clients will benefit the most from these technologies. Although the use of IoT in the production of renewable energy has many advantages, it is not without its issues and hurdles. One of the major issues confronting the usage of IoT in renewable energy is the initial investment expenditures. Despite significant price reductions in recent years, renewable energy technology remains prohibitively expensive. Improving the grid with many IoT sensors and data storage and monitoring equipment raises the initial investment expenses in renewable energy operations. Following the current trend, IoT-enabled energy production [8] equipment could gain widespread acceptance if prices fall dramatically in the future. The potential of hacking is another barrier to the use of IoT devices. Because the devices are linked to a network, a cyber-attack could occur if the network is not adequately secured. This can put utility providers and residents in dangerous and unfavorable positions. A safe procedure must be implemented to prevent any potential grid assault.

Renewable energy [9] sources are the way of the future in the production of power. According to the International Energy Agency, renewable energy is expected to grow by another 43% by 2022. Many businesses and governments are actively investing in smart cities worldwide, and smart energy technologies power these smart cities. Technologies such as computational vision, AI [10], profound training, strengthened learning, and intelligent energy enable the efficient operation and management of urban assets. It is a critical component in supplying renewable energy [11] to these smart cities. The United States, China, and India are the top markets for practices in renewable energy production. These marketplaces give enterprises working with IoT [12] services enormous business opportunities. IoT will play a vital role in increasing the dynamics of a clean energy source in renewable energy production. The inclusion of IoT in renewable energy offers both corporations and general customers enormous opportunities. To develop an IoT-focused learning model in renewable energy generation (REG) and energy conservation (Fig. 5.1).

We study and analysis different deep learning algorithm for use of IoT in REG for Conservation of Energy (CE) And proposed a hybrid deep learning (HDL) algorithm (AI [13] based on RL and LSTM).

In brief, the major aids of this research activities can be summarized as follows:

- A new agenda for simulation was proposed and a novel simulation explicitly designed for IoT sensor-based scenarios for testing energy-efficient models.

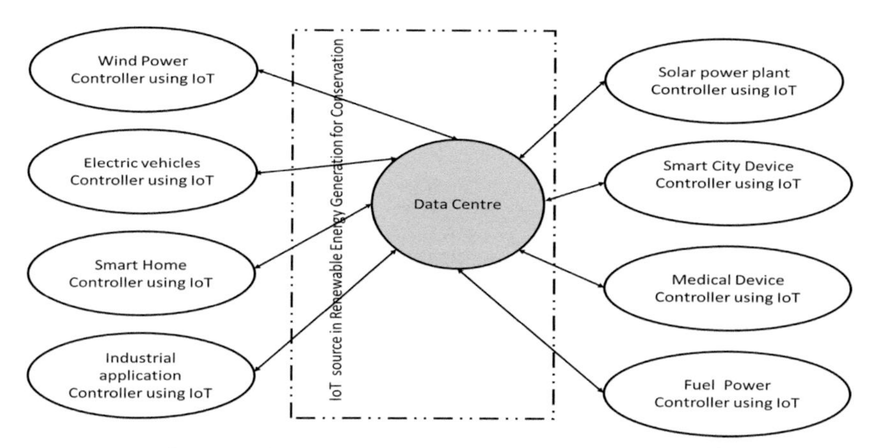

Figure 5.1 Application of internet of things in data centre for controlling internet of things devices in different domain.

- This study created a HDL neural utilization agent based on LSTM that learns resource utilization and produces better outcomes in a shorter training period than a typical deep-learning agent.

To organazie chapter Section 5.2 is concerned with the related work, Section 5.1 IoT and renewable energy Section 5.3 Proposed methodology, Section 5.4 Deep Q-learning which investigates the systems offered, provides the structure for our investigation. Section 5.5 Results analysis & discussion of the experiments.

5.2 Related work

The CE resources is vital, and artificial intelligence (AI) or machine learning [14] played an important role in fulfilling this purpose. The part of the IoT devices varies based on the nature of the functional usage of nodes. The function varies from the first node in the initial region to the last node of the network. The node's energy [15] directly affects the transmission coverage range of the IoT device. Most of the low-range communication IoT devices are connected in the sensing layer, which consumes minimum energy. Transmitting data from a low-range IoT device independently over a long distance [16] will require more energy and bandwidth. Reliable IoT in REG is needed to cover the entire region with minimal transmitting [17] nodes, conserving energy, bandwidth, and latency. We study and analysis reliable communication over the emerging

IoT environment, the following factors need to be considered. They are spectral efficiency, energy efficiency, connectivity, and REG for CE. We study the energy or resource constraint scenario for the sensing layer, a wireless paradigm is crucial. The coverage range of the IoT devices varies with their transmission distance, type of communication, and the type of data they perceive. They differ based on the spatial and temporal utilization of resources which directly influences the lifetime of networks. The best practice for energy conservation is developing a system that maximizes optimal efficiency [8].

The IoT sensors are used to monitor, evaluate the parameters, and compare results from the different platforms such as power stations, control rooms, and the system itself [9]. The design of Triboelectric nanogenerator is used for detecting real-time speed of wind energy due to rotatory motion of blades and it also used for monitoring purposes [10,11]. The sensors used for conservative energy environments should be reliable, durable, and able to work on flexible latency and bandwidth.

Additionally, the sensors should contain energy conversant and time-tolerant [12]. The leading technologies of AI are used in power systems where the traditional methods are not favorable due to the amount of energy and data produced [13]. Machine learning techniques are essential in generating results based on large raw database systems and dynamic processing [14]. The relationship between solar energy, wind energy, and coal production has been explained, and discuss the significant steps required to lessen the effect of harmful factors [15].

5.2.1 Internet of things and renewable energy

With the rapid advancement of Information Technology, a dynamic, connected IoT device has emerged, with things associated with mobile devices. Here, we provide the IoT for greater load control and REG for CE [18] and information on the state of the linked devices [19].

The electric power, wind power, smart home, solar power, smart city, water level, and current sensors [20] are among the sensors used for developing a robust system with a minimum error rate. This is accomplished through the use of sensors. LDR (light dependent resistor) exhibits photoconductivity, which reduces the resistance of the photosistor by increasing the intensity of the incident light. A IoT based [21] water level sensor prevents overflow and maintains tank balance. A small solar system can provide AC or DC electricity to a specific house or institution or even a particular piece of equipment. The turbine rotates at the speed of the wind and is linked to the DC generator in the Air Ventilator. The wind turbine rotates at less than or equal to 1 km/hour even at low speeds, delivering continuous power. In off-grid operation, mobile PV systems and fans can provide power based on consumer demand.

5.3 Proposed methodology

In a smart city, practical techniques of AI [22] can be employed with IoT enable smart home AI-driven algorithms to handle household energy use better. AI can operate with pattern reconnaissance approaches in a IoT enable [23] smart home. AI can utilize probabilistic and non-probabilistic designs in a smart home [24]. AI models are grouped into neurons, trees, Bayes, and supported vector machines. Models are used for AI models. All models of this kind need supervised training. Random field, Hidden Markov Models, and hidden Markov models are AI-bases of supervised models. In IoT enable smart home, artificial neural networks (ANNs) include nonprobabilistic models. Because neural networks may not consistently find the same solutions for a single input, neural network-based systems/solutions require a significant amount of training. Fig. 5.2 depicts the input signal processing pipeline used to estimate signal energy. The energy estimation [25] findings of the input signal are utilized [26] to conduct actions on smart home installed hardware/software-based features. Mobile phones can also be used to control electrical appliances using voice commands in a smart home. Users can also utilize their mobile phone's voice input, IoD device [27] and Bluetooth or Wi-Fi

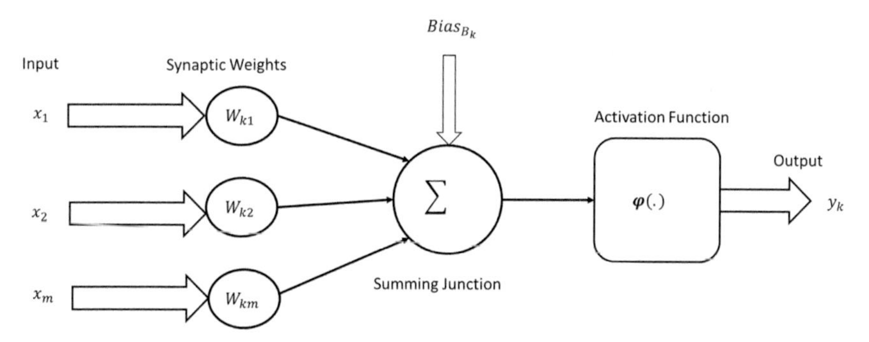

Figure 5.2 Mathematical representation of artificial neural network.

connections to share their voice with home energy management [28] systems and decode the action required on electrical items [29]. An ANN simulates human brain activities to research, resolve, and perform complex tasks.

In the Fig. 5.2, the contributions are denoted as x_1, x_2, x_3,. . .x_m. Before being transmitted to an activation feature, these inputs are multiplied by the weight of a binding to estimate the neuron output. The masses are denoted by, w_{k1}, w_{k2},. . .w_{km}. This operation is depicted mathematically in the following equation:

$$x_1 * w_{k1} + x_2 * w_{k2} + x_3 * w_{k3} = \sum (x_i * w_i) \tag{5.1}$$

(1) Applying the activation function gives, $\emptyset(\sum x_i{}^{*} w_i)$
(2) The activation function is a vitalmodule that regulates the overall network performance.

When recalling long-term settings, the LSTM [18] is a more complex type of reinforcement learning (RL) that is more reliable than regular RL. The LSTM's [30] gating mechanism, which enables the modeling of temporal sequences, is to blame. This chapter focuses on the type of LSTM layer illustrated in Fig. 5.1 that does not contain a peephole link.

Algorithm 1: There is a memory cell (played in green), three gates – one input I an output gate (o), and a gate that is forgotten (f) = element-wise multiplier. Functions for tangent activation: p, tanh = sigmoid and hyperbolic explained and the working of hybrid algorithm explained through this mathematical model is as follows:

$$v_j = (x_j, x_{j+1,...,} x_{j+k+1}) \tag{5.2}$$

Filter m association by the window vectors in the form of k-grams in each location in a mode to develop a feature map $E \in \mathrm{RL} - k + 1$; where each element E_j of feature mapped on window vector V_j is created as given:

$$E_j = f(V_j \times m + b) \tag{5.3}$$

$$W_j * V = (E_1, E_2, E_3, E_4 . . . E_n) \tag{5.4}$$

$$i_s = \sigma(v_j[\mathrm{HS}_{s-1}, x_s] + b_i) \tag{5.5}$$

$$f_s = \sigma(v_f[\mathrm{HS}_{s-1}, x_s] + b_f) \tag{5.6}$$

$$P_s = \tanh\text{HS}(v_j[\text{HS}_{s-1}, x_s] + b_P) \tag{5.7}$$

$$Q_s = \sigma(v_j[\text{HS}_{s-1}, x_s] + b_Q) \tag{5.8}$$

$$E_1 = f_s \varnothing E_{s-1} + i_s \varnothing P_s \tag{5.9}$$

$$\text{HS}_s = Q_s \tanh\text{HS}(E_1) \tag{5.10}$$

$$\text{HS}_p(P) = \sum_{i=1}^{N} y_i.\log(p(y_i) + (1 - y_i).\log(1 - p(y_i)) \tag{5.11}$$

$$\int_0^n \frac{\partial(\hat{\text{w}} * \hat{e})}{\partial x.\partial y} = J(x, y) \tag{5.12}$$

We have created derivatives in vector form for the sequential obtained by the LSTM algorithm. It generally forms a matrix of numerical values obtained by previous and used by hidden layers of neural networks. Similarly, Q_s is the set of sequences of such matrix available in memory space in queuing manner. In the E_1 is the first derivative of Q_s and it is used for obtaining HS_p, logarithmic function. Furthermore, we have integrated the hyper-space to obtain its derivative for n number of IoT device. Fog computing extends its services to the cloud by storing data locally in the fog node rather than broadcasting data to the cloud, which increases the pressure on the cloud. This improves cloud performance and increases fog computing efficiency. Fog computing reduces the rate at which data is processed and transferred to the cloud, allowing the cloud to retain and analyze data. This fog computing [31] service reduces the latency of network traffic. Fig. 5.3 depicts a graphical representation of fog computing. The fog computer is responsible for positioning the entire network fog node. Controller switch devices act as fog nodes and are then transferred to specified locations, such as electric power, wind power, smart home, soller power, smart city, and water level (Table 5.1).

5.4 Deep Q-learning

Q-learning is a model-free RL [32] algorithm is a an unsupervised machine learning algorithm for improving learning. The goal of Q-learning is used for IoT in REG for CE to create the agent's optimal

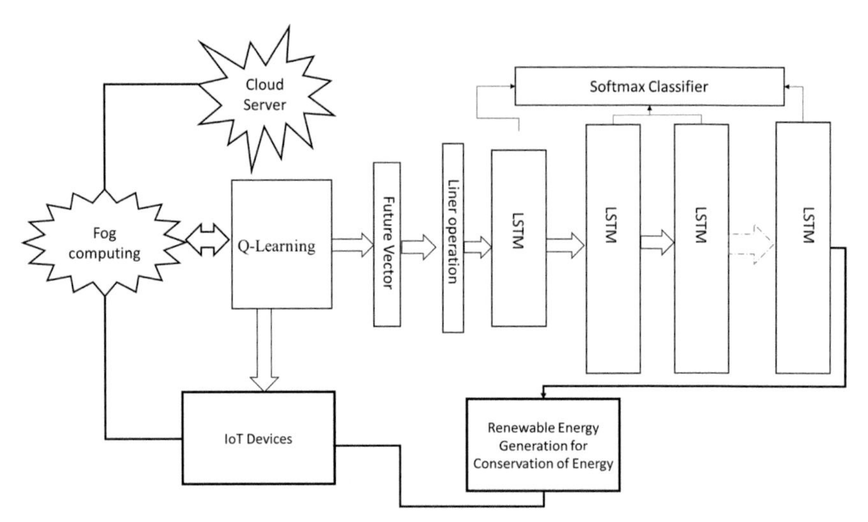

Figure 5.3 Flow-chart of interaction of Q-learning with different domains.

Table 5.1 Comparative analysis between cloud and fog computing for IoT device in renewable energy generation.

Parameter	Cloud computing Environment	Fog computing Environment
Delay	High delay—high energy	Low delay—low energy
Latency	High latency—high energy	Low latency—low energy
Mobility support	Low—low energy	Support—low energy
Geo distribution	Centralized—low energy	Supported—low energy

policy with the maximum reward to achieve the ultimate goal. This does not necessitate an atmospheric design and can handle transformations with shocks and incentives without requiring any changes. Q-learning determines the optimal approach for any finite Markov decision-making process [33]. To maximize the expected value of the most significant cumulative succession of prizes originating in today's state.As stated in the algorithm, this is a non radical learning algorithm that seeks the greatest possible consequence or movement under the condition. They are referred to as nonpolitical since the q-learning system detects decisions beyond existing policy, such as applying tiny random measures. Q-learning also works hard to implement a strategic approach that maximizes total motivation. This methodology's feedback system typically leads to enhanced learning accuracy and REG for CE.

Algorithm 2: Update Q-learning algorithm

Need:

Sate ST $= \{1, X_n\}$

Action AN $= \{1, \ldots X_n\}$as recognized AN:ST $= >$AN

Reward P: ST $\times$ Ą- $>$ ST

State transition (TS) TS:ST $\times$ Ą- $>$ ST

Learning rate(LR) $\in [0, 1]$

Specified $\alpha = 0.1$

Discounting factor $\gamma \in [0, 1]$

Generating Q Learning (ST, AN, TS, α, γ)

Modify Q learningSate $\times$ Ą $\rightarrow AR$

While Q learning is not joined do

$\sigma(N) = \text{argmax}\beta Q(N, \beta)$

$\beta < -\sigma(N)$

Received $\leftarrow L(ST, \beta)$

New state $\leftarrow TS(ST, \beta)$

Q learning (New state, $\beta) \leftarrow (1 - \alpha)$. Q learning $(ST, \beta) + \alpha$. (Received $+ \gamma$.)

Max Q learning (ST', β')

ST $\leftarrow ST'$

return Q

The primary goal of the study is to exploit IoT sensor resources in a closed smart sensor grid. As a result, the application was created specifically for use in the application mentioned above sector. The suggested AI-based RL with LSTM has the same structure as depicted in Fig. 5.3. The prior action's contribution, the state, and the reward will be used as input vectors. A later ensemble is a mix of several layers of LSTM with batch normalization to reduce unnecessary fit and model. A layer of batch normalization the following components was fed into a typical deep research layer known as the fully connected layer. The overall layer and node count vary depending on the amount of inputs from the simulation or the actual signal input of the circuit. One of these nodes then examines the data generated by IoT devices [34]. Data is evaluated in this fog-node without being sent back to the cloud, reducing cloud strain and minimizing latency and congestion for cloud activities for saving the energy. Cloud computing and fog computing differ from cloud computing in that they allow centralized access to resources. On the other side, fog computing provides decentralized access. Even though fog computing includes cloud computing capabilities, some things must be addressed to boost fog computing efficiency and efficiency as the number of IoT devices grows. The concept entails providing IoT devices with a one-of-a-kind storage space and computing capacity. The majority of IoT applications require low latency. In terms of latency, the cloud has an impact on their performance. Fog computing has been established to address this issue, where services that are away from the cloud can be offered, reducing network overcrowding and latency. Fig. 5.4 depicts the Four-level framework (Energy Generation, energy transmission, energy distribution, and energy consumption). Optimize IoT device operation and applications necessitate massive management capability for energy and data storage to ensure that broadband information is continuously provided dynamically and swiftly. For IoT-based applications, IoT and distributed computer integration are required. Cloud storage is a complete solution based on the required quality of service and a pay-as-you-go pricing mechanism [35]. It has a lot of storage space and computer skills.

However, some challenges have been demonstrated in fog computing for computing used for connecting all IoT device for distributed

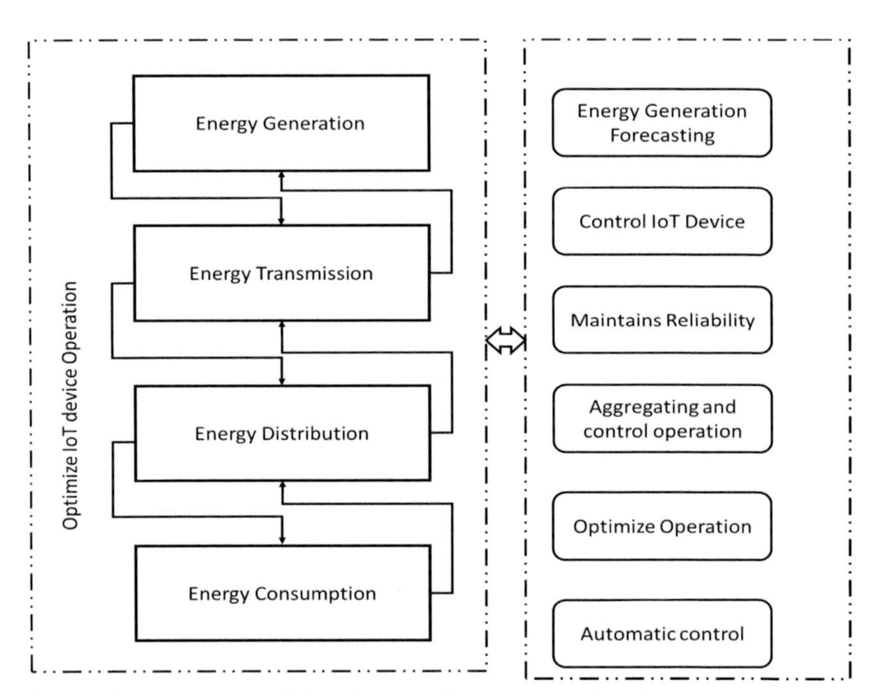

Figure 5.4 Current state of digitalization of the energy value chain.

envirment. The present study recognizes and highlights various research challenges related to fogs that must be addressed to provide a capable and visible resolution of developing services with a wide range of needs for the future of fog computing. One of the most critical issues is the timing of resource production and the avoidance of bottlenecks. As a result, the present study focuses on addressing some of the issues raised in the following sections.

Improved information availability throughout the value chain enables more vital decision support tools (e.g., AI) and remote control and automated decision-making (e.g., control of millions of devices with immediate actions, energy, such as algorithm trading or electric power, wind power, smart home, soller power, smart city, and water level). Digital monitoring and control technologies for energy saving have been used in energy generation and transmission for several decades and have recently begun to permeate more deeply into power systems.

IoT can improve asset and business management, resulting in more excellent reliability, higher safety, and new services and business models. This brief focuses on IoT apps that support VRE's high-share integration. The disruptive effects of sending and receiving massive amounts of granular data through networks are still being investigated in tens of billions of connected sensors and devices. Demand-side management, for example, can unleash greater flexibility in power systems. IoT also provides solutions for optimizing supply and demand systems, opening up enormous opportunities for increasing shares of REG to be integrated into the system. This would allow renewables to participate in and operate the system in electricity markets. Failure to manage transmission systems is expected to become more common as wind output levels rise. Distributed energy resources (DERs).

5.5 Results analysis and discussion

Techniques proposed for the proficient communication with IoT in REG for CE between devices employing hybrid learning in an IoT setting, LSTM, and proposed hybrid (Q-learning + LSTM). To put them into action, use the NS-3 simulator. The fitness element of the technology promotes dynamism by including the node's pace. The most efficient member was selected as the cluster leader, and the construction of the remaining node was found. The diagram depicts the various strategies and residual node formation (Fig. 5.5). In comparison to earlier methodologies such as LEACH [36], Fuzzy leach [37], HEED [38] and hybrid algorithm, the found strategy completely reduces residual nodes. The proposed hybrid algorithm (Algorithms 1 and 2) converts the formation of residual nodules into a separate cluster head and immediately transmits [39] data as even the node with the lowest appropriate fitness score.

The overall energy consumption of this method is proposed in Fig. 5.5 in comparison to other ways such as LEACH, Fuzzy leach, HEED and hybrid algorithm. The likelihood with respect to energy is also calculated in the Q-learning fitness value as well as in the decreasing in data transfer utilizing the hybrid algorithm. Because the most energy-efficient node is chosen at each level, energy consumption is reduced relatively.

The LSTM assists in reducing the dimensionality of the data sent to a minimum of $1/n$ dimensions of the data collected. This is a loss compression approach that is appropriate for all sizes. Fig. 5.6 depicts the network's lifetime when various techniques are used, which increases the accuracy [40] of reconstructed data. The network lifetime (NL) is the amount of time that the entire network is active or operational. The graphic in this case shows the network duration versus time in milliseconds.

The suggested system is compared to other approaches such as LEACH, Fuzzy leach, HEED and hybrid algorithm. The suggested system

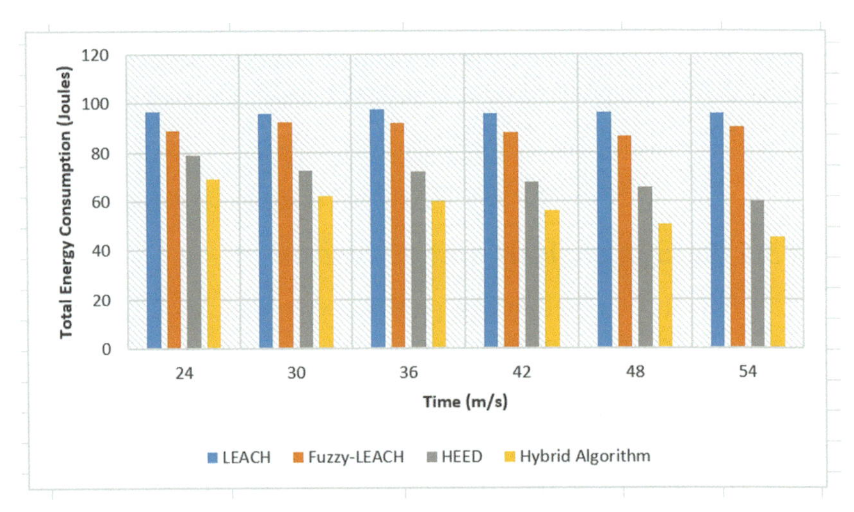

Figure 5.5 The total energy consumption at different time slots.

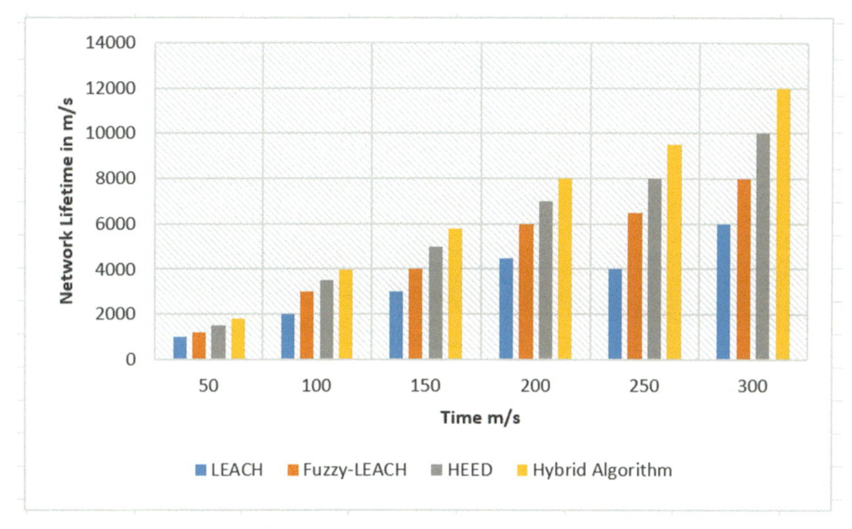

Figure 5.6 The network lifetime of the various approaches.

has a longer network life than the methods currently in use for dynamic-cluster creation employing the proposed algorithms such as Q-learning, LSTM for dimensionality reduction in a changeable environment, and fusion algorithms for reliable data transfer to base station (BS). All proposed approaches include IoT [41] node energy as one of their parameters. Since the lesser data transmission leads to lesser bandwidth consumptions, the proposed methods provide lesser bandwidth than the existing methods, but with tolerable data loss.

Fig. 5.6 The NL of the various approaches and since the lesser data transmission leads to lesser bandwidth consumptions, the proposed methods provide lesser bandwidth than the existing methods, but with tolerable data loss.

In the Fig. 5.7 compretive analysis different algorithm (LEACH, Fuzzy leach, HEED, and hybrid algorithm) between data transmission and time.

5.6 Conclusion and future work

A proficient communication system between IoT devices in a heterogeneous IoT environment, node sensor devices are proposed. The proposed learning-based reinforcement cluster training out performs the current IoT nodes technique. Q-learning and LSTM aids in consolidating

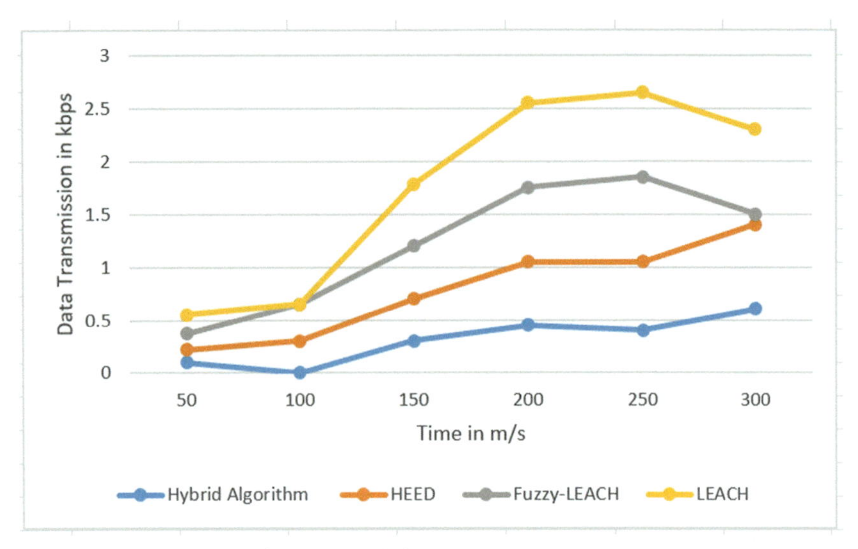

Figure 5.7 Comparison of transmitted data.

numerous data types, reducing the data dimension and the number of packets delivered, and boosting energy efficiency. The proposed Q-learning and LSTM also ensures the path to the base station's reliability based on the probability of the path with the highest energy, speed, and proximity value. The network's life is also extended through lowering electricity use. The system can be upgraded to increase the accuracy of the data recovery procedure at BS for any dynamic application by adding more sensor heterogeneity and an enhanced Q-learning and LSTM.

References

[1] W. Tushar, N. Wijerathne, W.T. Li, C. Yuen, H.V. Poor, T.K. Saha, et al., Internet of things for green building management: disruptive innovations through low-cost sensor technology and artificial intelligence, IEEE Signal Process. Mag. 35 (5) (2018) 100–110.

[2] S. Singh, P.K. Sharma, B. Yoon, M. Shojafar, G.H. Cho, I.H. Ra, Convergence of blockchain and artificial intelligence in IoT network for the sustainable smart city, Sustain. Cities Soc. 63 (2020) 102364.

[3] A.G. Matani, Internet of things and internet of drones in the renewable energy infrastructure towards energy optimization, in: R.N. Shaw, N. Mendis, S. Mekhilef, A. Ghosh (Eds.), AI and IOT in Renewable Energy. Studies in Infrastructure and Control, Springer, Singapore, 2021, pp. 15–26. Available from: https://doi.org/10.1007/978-981-16-1011-0_2.

[4] Y. Zhang, P. Geng, C.B. Sivaparthipan, B.A. Muthu, Big data and artificial intelligence based early risk warning system of fire hazard for smart cities, Sustain. Energy Technol. Assess. 45 (2021) 100986.

[5] S. Karad, R. Thakur, Efficient monitoring and control of wind energy conversion systems using Internet of things (IoT): a comprehensive review, Environ. Dev. Sustain. (2021). Available from: https://doi.org/10.1007/s10668-021-01267-6.

[6] M.A. Albreem, A.M. Sheikh, M.H. Alsharif, M. Jusoh, M.N. Mohd Yasin, Green internet of things (GIoT): applications, practices, awareness, and challenges, IEEE Access. 9 (2021) 38833–38858.

[7] T. Zhou, J. Shen, S. Ji, Y. Ren, L. Yan, Secure and intelligent energy data management scheme for smart IoT devices, Wirel. Commun. Mob. Comput. 2020 (2020). Available from: https://doi.org/10.1155/2020/8842885. Article ID 8842885, 11 pages.

[8] C. Tomazzoli, S. Scannapieco, M. Cristani, Internet of things and artificial intelligence enable energy efficiency, J. Ambient. Intell. Human. Comput. (2020). Available from: https://doi.org/10.1007/s12652-020-02151-3.

[9] P.P. Singh, P.K. Khosla, M. Mittal, Energy conservation in IoT-based smart home and its automation, in: M. Mittal, S. Tanwar, B. Agarwal, L. Goyal (Eds.), Energy Conservation for IoT Devices. Studies in Systems, Decision and Control, vol. 206, Springer, Singapore, 2019, pp. 155–177. Available from: https://doi.org/10.1007/978-981-13-7399-2_7.

[10] D. Kim, I.W. Tcho, Y.K. Choi, Triboelectric nanogenerator based on rolling motion of beads for harvesting wind energy as active wind speed sensor, Nano Energy 52 (2018) 256–263.

[11] Y. Xi, H. Guo, Y. Zi, X. Li, J. Wang, J. Deng, et al., Multifunctional TENG for blue energy scavenging and self-powered wind-speed sensor, Adv. Energy Mater. 7 (12) (2017) 1602397.

[12] C. Choi, J. Jeong, I. Lee, W. Park, LoRa based renewable energy monitoring system with open IoT platform, 2018 International Conference on Electronics, Information, and Communication (ICEIC), IEEE, 2018, pp. 1–2. Available from: https://doi.org/10.23919/ELINFOCOM.2018.8330550.

[13] H. Yousuf, A.Y. Zainal, M. Alshurideh, S.A. Salloum, Artificial intelligence models in power system analysis, in: A. Hassanien, R. Bhatnagar, A. Darwish (Eds.), Artificial Intelligence for Sustainable Development: Theory, Practice and Future Applications. Studies in Computational Intelligence, vol. 912, Springer, Cham, 2021, pp. 231–242. Available from: https://doi.org/10.1007/978-3-030-51920-9_12.

[14] C. Magazzino, M. Mele, G. Morelli, The relationship between renewable energy and economic growth in a time of COVID-19: a machine learning experiment on the Brazilian economy, Sustainability 13 (3) (2021) 1285.

[15] C. Magazzino, M. Mele, N. Schneider, A machine learning approach on the relationship among solar and wind energy production, coal consumption, GDP, and CO_2 emissions, Renew. Energy 167 (2021) 99–115.

[16] R. Anaadumba, Q. Liu, B.D. Marah, et al., A renewable energy forecasting and control approach to secured edge-level efficiency in a distributed micro-grid, Cybersecur 4 (2021) 1. Available from: https://doi.org/10.1186/s42400-020-00065-3.

[17] C. Anitescu, E. Atroshchenko, N. Alajlan, T. Rabczuk, Artificial neural network methods for the solution of second order boundary value problems, Comput. Mater. Cont. 59 (1) (2019) 345–359.

[18] E. Azari, S. Vrudhula, An energy-efficient reconfigurable LSTM accelerator for natural language processing, 2019 IEEE International Conference on Big Data (Big Data), IEEE, 2019, pp. 4450–4459. Available from: https://doi.org/10.1109/BigData47090.2019.9006030.

[19] B. Yan, F. Hao, X. Meng, When artificial intelligence meets building energy efficiency: a review focusing on zero energy building, Artif. Intell. Rev. 54 (2021) 2193–2220. Available from: https://doi.org/10.1007/s10462-020-09902-w.

[20] C. Kim, B.G. Kim, J.J.P.C. Rodrigues, Editorial: recent advances in mining intelligence and context-awareness on IoT-based platforms, Mobile Netw. Appl. 24 (2019) 160–162. Available from: https://doi.org/10.1007/s11036-018-1172-2.
[21] M. Srbinovska, V. Dimcev, C. Gavrovski, Energy consumption estimation of wireless sensor networks in greenhouse crop production, IEEE EUROCON 2017 – 17th International Conference on Smart Technologies, IEEE, 2017, pp. 870–875. Available from: https://doi.org/10.1109/EUROCON.2017.8011.
[22] A.S.H. Abdul-Qawy, T. Srinivasulu, SEES: a scalable and energy-efficient scheme for green IoT-based heterogeneous wireless nodes, J. Ambient. Intell. Human. Comput. 10 (2019) 1571–1596. Available from: https://doi.org/10.1007/s12652-018-0758-7.
[23] F.M. Al-Turjman, Information-centric sensor networks for cognitive IoT: an overview, Ann. Telecommun. 72 (2017) 3–18. Available from: https://doi.org/10.1007/s12243-016-0533-8.
[24] S. Bansal, D. Kumar, IoT ecosystem: a survey on devices, gateways, operating systems, middleware and communication, Int. J. Wireless. Inf. Networks. 27 (2020) 340–364. Available from: https://doi.org/10.1007/s10776-020-00483-7.
[25] S. AbbasianDehkordi, K. Farajzadeh, J. Rezazadeh, et al., A survey on data aggregation techniques in IoT sensor networks, Wireless Netw. 26 (2020) 1243–1263. Available from: https://doi.org/10.1007/s11276-019-02142-z.
[26] E. Suganya, C. Rajan, An adaboost-modified classifier using particle swarm optimization and stochastic diffusion search in wireless IoT networks, Wireless Netw. 27 (2021) 2287–2299. Available from: https://doi.org/10.1007/s11276-020-02504-y.
[27] N. Hossein Motlagh, M. Mohammadrezaei, J. Hunt, B. Zakeri, Internet of things (IoT) and the energy sector, Energies 13 (2) (2020) 494. Available from: https://doi.org/10.3390/en13020494.
[28] Y. Belkhier, A. Achour, R.N. Shaw, A. Ghosh, Performance improvement for PMSG tidal power conversion system with fuzzy gain supervisor passivity-based current control, in: S. Mekhilef, M. Favorskaya, R.K. Pandey, R.N. Shaw (Eds.), Innovations in Electrical and Electronic Engineering. Lecture Notes in Electrical Engineering, vol. 756, Springer, Singapore, 2021. Available from: https://doi.org/10.1007/978-981-16-0749-3_6.
[29] G. Kapoor, V.K. Mishra, R.N. Shaw, A. Ghosh, Fault detection in power transmission system using reverse biorthogonal wavelet, in: S. Mekhilef, M. Favorskaya, R.K. Pandey, R.N. Shaw (Eds.), Innovations in Electrical and Electronic Engineering. Lecture Notes in Electrical Engineering, vol. 756, Springer, Singapore, 2021. Available from: https://doi.org/10.1007/978-981-16-0749-3_28.
[30] M. Konstantinou, S. Peratikou, A.G. Charalambides, Solar photovoltaic forecasting of power output using LSTM networks, Atmosphere 12 (2021) 124. Available from: https://doi.org/10.3390/atmos12010124.
[31] P. Singh, S. Bhardwaj, S. Dixit, R.N. Shaw, A. Ghosh, Development of prediction models to determine compressive strength and workability of sustainable concrete with ANN, in: S. Mekhilef, M. Favorskaya, R.K. Pandey, R.N. Shaw (Eds.), Innovations in Electrical and Electronic Engineering. Lecture Notes in Electrical Engineering, vol. 756, Springer, Singapore, 2021. Available from: https://doi.org/10.1007/978-981-16-0749-3_59.
[32] M. Kumar, V.M. Shenbagaraman, R.N. Shaw, A. Ghosh, Predictive data analysis for energy management of a smart factory leading to sustainability, in: M.N. Favorskaya, S. Mekhilef, R.K. Pandey, N. Singh (Eds.), Innovations in Electrical and Electronic Engineering. Lecture Notes in Electrical Engineering, vol. 661, Springer, Singapore, 2021, pp. 765–773. Available from: https://doi.org/10.1007/978-981-15-4692-1_58.

[33] G. Kapoor, P. Walde, R.N. Shaw, A. Ghosh, HWT-DCDI-based approach for fault identification in six-phase power transmission network, in: S. Mekhilef, M. Favorskaya, R.K. Pandey, R.N. Shaw (Eds.), Innovations in Electrical and Electronic Engineering. Lecture Notes in Electrical Engineering, vol. 756, Springer, Singapore, 2021, pp. 395–407. Available from: https://doi.org/10.1007/978-981-16-0749-3_29.
[34] R.N. Shaw, P. Walde, A. Ghosh, IOT Based MPPT for performance improvement of solar PV arrays operating under partial shade dispersion, 2020 IEEE 9th Power India International Conference (PIICON), Deenbandhu Chhotu Ram University of Science and Technology, Sonepat, India, February 28–March 1, 2020.
[35] H.K. Huneria, P. Yadav, R.N. Shaw, D. Saravanan, A. Ghosh, AI and IOT-based model for photovoltaic power generation, in: S. Mekhilef, M. Favorskaya, R.K. Pandey, R.N. Shaw (Eds.), Innovations in Electrical and Electronic Engineering. Lecture Notes in Electrical Engineering, vol. 756, Springer, Singapore, 2021, pp. 697–706. Available from: https://doi.org/10.1007/978-981-16-0749-3_55.
[36] Y. Belkhier, A. Achour, R.N. Shaw, W. Sahraoui, A. Ghosh, Adaptive linear feedback energy-based backstepping and PID control strategy for PMSG driven by a grid-connected wind turbine, in: S. Mekhilef, M. Favorskaya, R.K. Pandey, R.N. Shaw (Eds.), Innovations in Electrical and Electronic Engineering. Lecture Notes in Electrical Engineering, vol. 756, Springer, Singapore, 2021, pp. 177–189. Available from: https://doi.org/10.1007/978-981-16-0749-3_13.
[37] Y. Yuan, T. Zhang, B. Shen, X. Yan, T. Long, A fuzzy logic energy management strategy for a photovoltaic/diesel/battery hybrid ship based on experimental database, Energies 11 (9) (2018) 2211. Available from: https://doi.org/10.3390/en11092211.
[38] R. Pawlak, B. Wojciechowski, M. Nikodem, New simplified HEED algorithm for wireless sensor networks, in: A. Kwiecień, P. Gaj, P. Stera (Eds.), Computer Networks. CN 2010. Communicaions in Computer and Information Science, vol. 79, Springer, Berlin, Heidelberg, 2010, pp. 332–341. Available from: https://doi.org/10.1007/978-3-642-13861-4_35.
[39] A.S. Rajawat, R. Rawat, K. Barhanpurkar, R.N. Shaw, A. Ghosh, Vulnerability analysis at industrial internet of things platform on dark web network using computational intelligence, in: J.C. Bansal, M. Paprzycki, M. Bianchini, S. Das (Eds.), Computationally Intelligent Systems and Their Applications. Studies in Computational Intelligence, vol. 950, Springer, Singapore, 2021, pp. 39–51. Available from: https://doi.org/10.1007/978-981-16-0407-2_4.
[40] A.S. Rajawat, R. Rawat, K. Barhanpurkar, R.N. Shaw, A. Ghosh, Blockchain-based model for expanding IoT device data security, in: J.C. Bansal, L.C.C. Fung, M. Simic, A. Ghosh (Eds.), Advances in Applications of Data-Driven Computing. Advances in Intelligent Systems and Computing, vol. 1319, Springer, Singapore, 2021, pp. 61–71. Available from: https://doi.org/10.1007/978-981-33-6919-1_5.
[41] A.S. Rajawat, K. Barhanpurkar, R.N. Shaw, A. Ghosh, Risk detection in wireless body sensor networks for health monitoring using hybrid deep learning, in: S. Mekhilef, M. Favorskaya, R.K. Pandey, R.N. Shaw (Eds.), Innovations in Electrical and Electronic Engineering. Lecture Notes in Electrical Engineering, vol. 756, Springer, Singapore, 2021, pp. 683–696. Available from: https://doi.org/10.1007/978-981-16-0749-3_54.

CHAPTER SIX

Renewable energy system for industrial internet of things model using fusion-AI

Anand Singh Rajawat[1], Omair Mohammed[2], Rabindra Nath Shaw[3] and Ankush Ghosh[4]

[1]Department of Computer Science Engineering, Shri Vaishnav Vidyapeeth Vishwavidyalaya, Indore, India
[2]University of the Cumberlands, Williamsburg, KY, United States
[3]Department of International Relations, Bharath Institute of Higher Education and Research (Deemed to be University), Chennai, India
[4]School of Engineering and Applied Sciences, The Neotia University, Sarisha, India

6.1 Introduction

Energy plays a crucial role in the global economic environment. The vital infrastructure includes electricity systems, which play an essential role in human life by supplying energy to meet home and industrial needs. Because power systems have human lives and expensive infrastructures such as turbines and generators, their automation mission becomes vital. According to the global energy perspective, roughly one-eighth of the world's population does not have access to electricity [1]. The major drivers of change include decarbonization, reliability with increased demand and transport electrification, customers with empowerment, market designs, and regulatory paradigms. The variables that make it feasible are advanced technology, policies, and standards [2]. Generation, transmission, distribution, and consumption are all factors in the supply chain. The transition from a centralized to a more distributed architecture is readily apparent. Turbines, generators, circuit breakers, switches, transformers, and other essential equipment are part of the industrial internet of things (IIoT) supply chain. Power system control, sensors [3], communication switches, monitoring and data acquisition, and distributed control systems are examples of secondary equipment used in automation. These measures are based on the voltage and current measurements' protection functionalities. Microsecond reaction times are required in power systems to isolate the defective portion using open-loop control.

Applications of AI and IOT in Renewable Energy.
DOI: https://doi.org/10.1016/B978-0-323-91699-8.00006-1

Closed-loop control with feedback is included in the controls. This only takes a few milliseconds. In the Industrial Age, energy efficiency [4], power generation, and sustainability are critical 4.0 [5]. Energy suppliers that are sensitive to rising usage of resources and scarcity and the environment are focusing on sustainability. Developed countries such as India and China have growth rates of up to 25%. Electrifying energy distribution, growing energy consumption, increasing dynamic electrical charges, Renewable Energy, and increasing dispersed power are the four essential themes. This research work examines the first landscape of Renewable energy systems (RESs) and their future evolution and future developments. The IoT is a critical component in the Industry 4.0 [6] shift. Intelligent factories can leverage IIoT technology to improve worker efficiency, safety, and output by integrating it into a wide range of machinery.

6.1.1 Renewable energy system for smart production

It is feasible to communicate and construct a linked plant by integrating intelligent technologies into production machines and processes [7]. Employees can access all data collected from this connected gadget in a central database. Managers can get a complete picture of the facility by storing data in one place. They can observe when specific machinery parts were in use, as well as details regarding their performance. They can also learn how many resources areconsumed, Energy consumed, how many units are created, and other information.

6.1.2 Energy management for renewable energy system

Because industrial operations are frequently energy-intensive [8], even little improvements in energy management can help industrial enterprises save considerable money while also improving their environmental performance. With IIoT, managers can better understand a company and facility's energy usage for RESs. Each machine's energy use can also be seen in detail. Companies can use data from intelligent sensors to optimize their energy use. It is possible to discover which machines consume the most energy and which machines are inefficient. It can even assist in the addition of renewables to a system.

6.1.3 Predictive maintenance

Industrial organizations can benefit from IIoT by improving their predictive maintenance programs and increasing their efficiency. The monitoring of equipment condition and performance to predict failure is known as predictive maintenance. The worker can avoid these failures by performing routine maintenance. Reactive maintenance, on the other hand, is when personnel fixes a machine after it breaks down. Intelligent sensors can provide specific information on how a machine component operates, allowing users to detect faults sooner and more precisely. Workers may be alerted to odd operations or potential issues. IoT-enabled predictive maintenance [9] helps businesses decrease downtime and increase productivity save the energy for RESs.

6.2 Related work

Puri et al. [10] this research work uses several sensors to construct a system that uses residential appliances and industrial places. The power generating circuit connects a variety of sensors, including a piezoelectric sensor, heat from the body, and a solar panel. Two artificial intelligence (AI) models, the artificial neural network (ANN) and the adaptive fuzzy inference system, are used to calculate the total energy generated from renewable energy resources (RERs).

Bedi et al. [11] by replicating biological nerve systems, both on the fringe and on the device, with cognitive calculation, streaming, and distributed analytics, the development of computer intelligence skills can produce an intelligent IoT system. The electrical and energy transformation systems are examined in this review study.

Sherazi et al. [12] across various sensor intervals is the main result of these surveys. This shows a linear increase in the total cost of £1500 in the industrial environment (nonenergy harvesting) over 5 minutes. The EH scenario tends to reduce up to 5 times after a given interval. Furthermore, carbon emissions from LoRa motes were measured at up to 3 kg/kWh in renewable sources, with yearly CO_2 emissions savings of up to 3 kg/kWh. The findings of this study could be crucial for a green business that strives for cost and energy efficiency.

Islam et al. [8] the electrical supply to specific residential loads was revealed in this study, to lower energy expenditures owing to environmental consequences.

Clairand et al. [13] energy efficiency methods are critical for this particular industrial sector. The different energy efficiency potentials in the food business are explored in this article—first, a quick rundown of the primary food industry and their respective energy use. The various energy efficiency options for thermal and electrical energy are then discussed. Industry 4.0 and demand response have also spawned new trends and opportunities.

Tuttokmaği et al. [14] fourth Industrial Revolution. Optimization, automation, energy efficiency, smart production, and the internet are all prevalent themes in the Industrial Revolution and Smart Grids. This research looks at how these process smart grids interact with one other and the process of building the Industrial 4.0 revolution globally and in our own country.

Ke et al. [15] optimize data transfer delays, energy consumption, and bandwidth allocation simultaneously while avoiding the dimensionality curse caused by a complex action space. By learning from the dynamic IoT environment, JODRBRL can reduce the entire cost of the system, including the cost of data buffer delays, energy usage, and bandwidth.

Hossain et al. [16] long speed long short-term memory (LSTM) and a fully connected neural grid are all part of the hybrid algorithm. Sophisticated models such as the multilayer neural network (NN).

6.3 Internet of things in renewable energy sector

Renewable energy sources will, without a doubt, be the dominant sources of energy in our future. Their adoption has been continuously expanding, allowing for the development of intelligent energy solutions. These resources are unquestionably preferable to those that release hazardous gases, which are already scarce. Renewable energy [17] is being deployed at a breakneck pace that is also cost-effective. Modern disruptive technologies [18] can help to improve the use of RERs. The use of IoT in the renewable energy sector has significantly increased. IoT applications helps to overcome several barriers to renewable energy adoption. Here are a few examples of renewable energy IoT applications that contribute to a more sustainable future:

6.3.1 Automation to advance complete production

Solar and wind energy are the most widely used renewable energy sources [19]. They have contributed to their quantity and dependability more than any other renewable energy source. In 2019 windmill farms provided enough energy to meet a fifth of Germany's energy consumption. Energy costs related to the creation of these resources [19] have decreased significantly. Solar panel prices have dropped by 99% since 1977. Germany, China, and Japan are the world's top solar energy producers. The integration of IoT with sensors in solar and wind systems can improve their dependability even more. To ensure optimal energy production efficiency. Analytical solutions [20] can be used to track the sun's movement, and the angle of the solar panels can be altered automatically. In the realm of wind energy, IoT [21] may be used to track various metrics related to electricity generation.

6.3.2 Smart grids for elevated renewable implementation

Traditional power grids cannot encourage this reliance on renewables based on weather conditions. Smart grids have been built due to the IoT [22], allowing for manual power disruptions between renewables and established power plants. Smart grids have aided in supporting the various characteristics of renewable energy and ensuring a continuous supply of energy to users.

6.3.3 The internet of things is increasing renewable energy adoption

The IIoT-enabled [23] development of intelligent grids has boosted the rise of renewable energy sources. They provide tremendous advantages in energy consumption monitoring and real-time notifications, allowing energy utilities to integrate renewable energy sources. The following are a few of these advantages:

6.3.3.1 Energy expenditures

End-users are now using renewable energy to lower their energy expenditures [24] and become self-sufficient. Many countries, such as India, provide solar subsidies to their residents to boost the usage of renewable energy. Countries assist residents in constructing solar plants on their rooftops for personal energy requirements.

6.3.3.2 Balancing supply and demand

Energy utilities use IIoT smart grids to ensure that consumers have a consistent supply of power. Integrating IoT [25] with renewable energy allows suppliers to accept renewable energy while also meeting the needs of end customers. Intelligent energy meters are being used commercially to supply electricity suppliers with real-time consumption statistics. They can also use analytics and data processing tools to establish trends and patterns related to high-load scenarios. As a result, utilities can utilize the manual switching technique to limit the utilization of power plants during regular off-peak times and then run them during times of excessive electricity demand. Utilities can thus manage supply and demand while also limiting dangerous material emissions into the environment.

6.3.3.3 Cost-effectiveness

According to studies, if only 12% of Saharan solar energy can be used, the world's energy needs can be supplied (around 110,400 km^2). However, there are still a few drawbacks to this strategy. It is challenging to create and operate a large solar farm [26], for example. Furthermore, electricity from this remote area will be subject to transmission and distribution losses. Power losses in transmission lines, for example, can approach 10% over long distances. These obstacles and stumbling blocks are impeding the expansion of solar power and renewable energy in general. The usage of IoTs in solar energy could help to minimize construction costs and solar station management. The real-time monitoring and forecasting analytics capabilities of the IIoT can be utilized to keep track on characteristics that can reduce power plant efficiency or result in unforeseen failures. As a result, businesses can save money on inspections and repairs while increasing efficiency.

6.4 Proposed methodology

IIoT can establish a comfortable working atmosphere within the building using IoT smart industrial environment sensors. Temperature and moisture sensors, leak and water sensors, air and air smoke sensors, and light sensors are among the IoT's [27] intelligent environment sensors. Humidity and temperature sensors: these sensors keep track of unanticipated variations in the temperature. Temperature and moisture sensors are also used to save energy by turning off refrigeration and heating while no one is present. Heating sensors, thermometers, thermocouples, infrared sensors, bimetallic devices, silicon diodes, and state change sensors are only a few examples. The following are some examples of temperature sensors: Humidity sensors can include resistive sensors and capacitive sensors. Leak and water sensors: Sensors warn homeowners to a leak, preventing costly floods. Water and leak sensors under carpet leak sensors, rope-like sensors, and hydroscopic tape-based sensors are examples of leakage sensors and water sensors. Leak and water sensors are two examples. Smoke and air sensors: these sensors keep track of indoor air quality. Smoke and air sensors allow IIoT workers to detect smoke, carbon monoxide, and any other harmful gas in their industry. As a result, industrial employees will take corrective action [28] before anyone in the industry suffers major harm. Smoke and air sensors include photoelectrical, sensor ionization, dual, inhalation and vapor sensors, projected beam sensors, video sensors, and heat sensors, to name a few.

6.4.1 Interruption attacks

A hardware or sabotage DoS attack and a software-based DoS attack are examples of interruption attacks. denial of service attack (1) counter-sabotage and sabotage (2) network device disconnection by methods of IoT [29] device hardware or infrastructure sabotage (e.g., cable cutting or damage caused to the physical IoT device). Cutting the link between PMU and PDCs and Super PDCs, or causing a physically damaged connection to PMUs and intelligent meters, are examples of EPES sabotage. Sabotage of the RES is another example of sabotage. Electric power plants are equally susceptible to sabotage. Attacks on vital IoT infrastructure can be reduced if access to it is restricted. Execute attacks and counter-attacks During a DoS attack, the attacker hacks several workstations (or zombies) and consumes network resources, overloading the target bandwidth and slowing or stopping legitimate traffic (also known as DDoS).

DoS attacks, for example, are causing delayed and lost measurements from devices that rely on real-time measurement data on RERs (e.g., PMUs and smart meters). As a result, the transmission system [30] condition, the delayed resolution of power system problems, or the complete failure of network measurement equipment cannot be reliably forecasted. Network layer assaults, transportation layer attacks, denial of service attacks on the Local Area Network, and teardrop attacks are all examples of DoS attacks. During teardrop attacks, attackers send fragmented packets to a target. Due to a flaw in reassembling the TCP/IP fragmentation, the target cannot reassemble the received packets, resulting in overlapping packets that crash the target network device. DoS attacks can severely harm the IIoT. As a result, network security mechanisms like air gaps, anomaly detection, huge pipes, and traffic filtering must be used less frequently. An airtight network is a network security solution that physically isolates a secure computer network from other dangerous networks public Internet or an insecure local area network. It disables nonlocal segment machine connectivity. The expensive expense of constructing separate network infrastructures, for the other hand, is a drawback to this method. Network DoS attacks are detected using anomaly detection techniques. Experiments revealed [31] that detection performance is inverted concerning network usage. The utilization of the network is also essential in determining the best detection parameters. Significant bandwidth connections are networks capable of absorbing attacks to mitigate DoS attacks. The high expenditures connected with this approach, on the other hand, constitute a disadvantage. Filtering traffic is a less expensive technique to protect networks against DoS assaults [32]. This approach employs a distributed or redundant infrastructure to reroute attack flow. This technique, however, has significant flaws, including a lack of documentation to back up the assertion that DoS traffic is filtered from ordinary traffic and the difficulties in implementing it.

The economic impact of the IoT on energy and electrical systems McKinsey For energy and power systems, While proposed approach offer huge revenue prospects, these outstanding numbers must be evaluated against the high costs associated with implementing new IoT devices [33] and technology. The surplus generated, on the other hand, exceeds the starting costs. In addition, IoT technologies can be used to save money on current devices and IIoT infrastructure. Connectivity, cyber security [34], massive data management, personal privacy, low-cost, sustainable power resources, or dependable sensors are now preventing IoT devices and technology from entering the market. To overcome these obstacles and ensure that the IoT for IIoT continues to grow, viable solutions are required. Energy and electrical power systems the IoT has an impact on the environment. In IIoT, power is used more efficiently with IoT. Control systems [35] have also been tweaked to maximize the use of renewable energy sources (solar and wind). This has a favorable influence on the environment by reducing energy waste and carbon dioxide (CO_2) emissions. Annual CO_2 emissions are predicted to drop by 2 gigatons by 2020. The social influence of the IoT on energy systems and electricity As the world's population grows, it becomes increasingly important for individuals to look after the planet's resources. Health, comfort, and convenience are increasingly personal priorities around the world as living standards rise. IoT can meet all of these demands and wants by sensing, gathering, transmitting, analyzing, and sharing large data. Organizations and institutions will use the IIoT to boost energy efficiency, control, and audit skills to achieve these standards. When hacking the smart meter, more personally identifiable data gathered from the meter could jeopardize personal security (e.g., data on energy use and user movements and activities to track data). A hacker, for example, could be able to tell whether or not a user is at home or whether or not a child is there. While IoT deployment in RES [36] for IIoT Model Using Fusion-AI poses a cyber and privacy risk, numerous social benefits include improved lifestyles, public safety, energy conservation, cost savings, and a healthier environment. Our proposed system also hasmany advantages. Individuals and businesses must decide on the best use of technology for their requirements based on these negotiating agreements—inability to push IoT adoption into society and gain widespread acceptance. People enjoy being in charge of their own well-being. Given the numerous benefits of deploying IoT technology, many people may be eager to give it a shot. Even if we are aware of the benefits, some people will be resistant to this technology. Furthermore, international rivalry for excellence in the manufacturing and development of IoT devices [37] makes it difficult for a company to establish a foreign base and deploy its resources. People's choices should be respected in all instances, and they should not be coerced onto a road that makes them uncomfortable.

There are several benefits to integrating IoT with RESs but there are also several obstacles. These problems include sensing, networking, power management, massive data, computers, complications, and safety [38]. Technical advances are required to meet these problems and produce an intelligent cyber-secured electric power network for RESs.

Renewable Energy Sources Integration: All through the last decades, the incorporation of renewable sources, mainly photovoltaic (PV) (solar PV system) are setting up and wind energy plants have led to a considerable dynamic characteristic changer in power systems. This alteration is mainly because the majority of renewable have different characteristics, availability/ certainty on energy. To integrate a huge sum of renewable energies into the open power system network, reconfigure the existing energy systems dynamics. The distributed nature of renewable sources

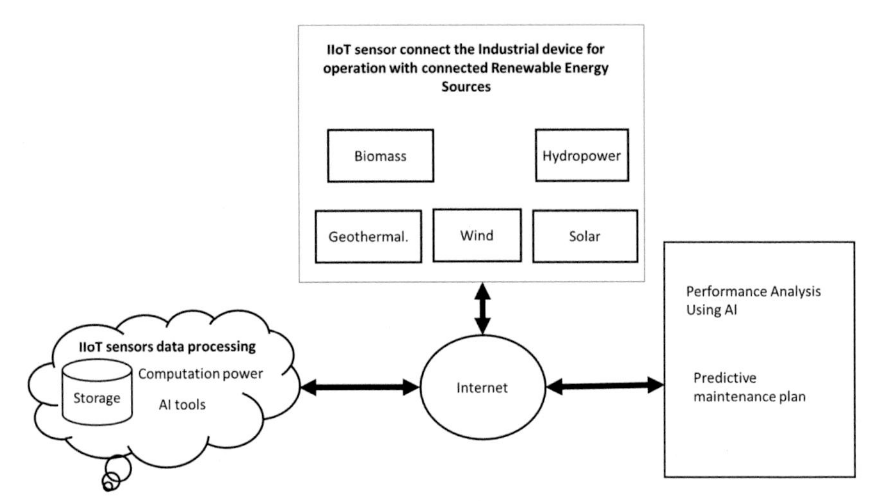

Figure 6.1 Renewable energy system for industrial internet of things model using fusion-AI.

(i.e., wind power, PV, biomass, fuel cell, etc.) is the future of reasonably priced power systems (Fig. 6.1).

6.5 Renewable energy system for industrial internet of things model

The difficulties are especially apparent in distribution networks when planning, operation, and maintenance are challenging. Integration of renewable energy sources and electric vehicles, energy storage, demand-side management, lighting and propulsion equipment, and response to recent extreme weather occurrences are examples of this. The services emphasize customer service, system dependability, and operational resiliency. Energy storage could provide some load support. This necessitates metering equipment, system monitoring, and control that is more modern and IIoT [39] protection systems. These are the difficulties:

1. Solar PV Rooftop with Distributed Resources Integration (DRI) is becoming increasingly popular. Wind turbines and fuel cells are two more prominent choices. It has the option of being connected to the grid or not. Utility-owned DRIs can be set to have the least amount of effect. DRIs are largely unpatched and has no impact on system

capability. These are not restricted in terms of frequency [40], but they continue to contribute to capacity deferment.

2. Energy storage planning and implementation There are chemical, thermal, and mechanical techniques. Energy storage separates energy production and delivery in minutes, hours, and days. Load shaping, maximum load delay, addition/backup power, power arbitration, power control, and frequency control are some benefits.
3. Transmission Systems Interfacing historically, two separate systems have been conveyed and distributed. With the bidirectional power flow, the lines are blurring. Between medium voltage and low voltage (LV), converging lines form (LV).
4. Natural factors and their effects on Tsunamis, monkey devices, and blackouts are examples of extreme weather disasters that destroy equipment. Modern weather forecasting allows operators and planners to plan for and rebuild damage avoidance infrastructure. By building redundant/direct communication paths beneath the feed system, circuits beneath the feed system help to improve system dependability and minimize service interruptions.

Microgrids [41] are one method of implementing distributed control, and autonomy is also discussed. Controlled and coordinated electricity distribution systems incorporate loads and dispersed energies that can be operated on an electricity grid or isolated. Microgrids [42] are similar to hybrid cars in that they are partially powered by electricity and have battery-shaped storage to ensure a constant supply. Because renewable sources, such as the sun, are intermittent, a reliable, nonenvironmentally friendly source must be included. The necessity for massive sensing capabilities, a safe and reliable communication system, Standard Interoperability, data management, viewing, archiving, data detection unsuitable, time synchronization, and data security and sharing are among the lessons learned from deployments around the world [43]. Only 20% of respondents say data might be used in industrial processes. According to McKinsey, It's also vital to have distributed intelligence. For distribution and privacy, emerging technologies necessitate cybersecurity.

In this work, we used a four-step methodology: (1) gather and process data, (2) send data to the cloud server, (3) develop the model, and (4) validate the model using the Fusion AI ANN. Fig. 6.2 depicts the proposed IoT model.

A piezoelectric module, electric body heating, and solar panels for sustainable energy are among the three modules. For the first component,

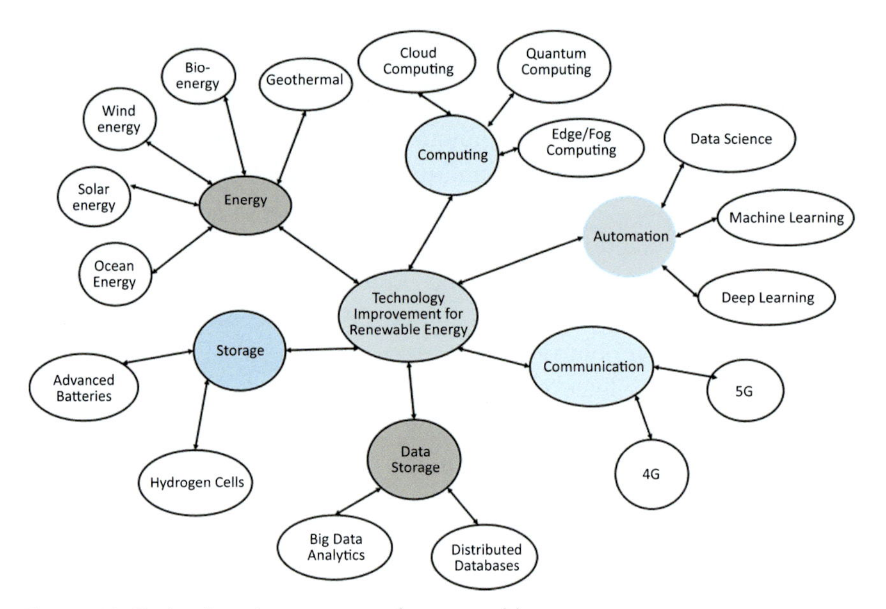

Figure 6.2 Technology improvement for renewable energy.

sensors can be deployed anywhere in the world (Fig. 6.2). Working of some IIoT device work for Mechanical tension is turned into the Working of some IIoT device electrical energy and output are pressure and pressure in the sensor.

The second component uses the heat produced by the Working of some IIoT device. This is monitored by the sensors, which send heat to the converter and the electrical producers. The sensor is coupled to a power storage device, which stores the electrical energy produced by the sensor. The energy is stored in the store and used for various things, including mobile charging and headlamps. The equation for this component is as follows: where is the Stefan-Boltzman constant (0174 Btu/hour-ft.^{2}−oR4), and is the heat transfer rate (Btu/hour) (oR). For the third component, photovoltaic cells are used (Fig. 6.3). When sunlight strikes photovoltaic cells, the electron loses its atom instantaneously. The photovoltaic cell has both a positive and negative side, making it a medium that flows through and generates energy. Solar panels are only capable of producing DC power. The circuitry, which consists of a diode and a regulator, stores DC power straight into batteries. Converting the DC power supply to an AC power supply is required for AC. Solar Panel frequently makes approximately 12 V and current differs rendering to the size of solar panel.

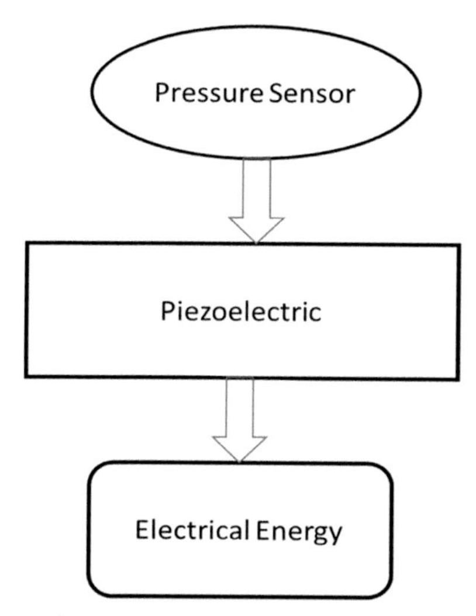

Figure 6.3 Working piezoelectric sensors.

The equation represents the solar power energy calculation

$$
\begin{aligned}
&\mathrm{EG} = \mathrm{Area} \times n \times \mathrm{Av} \times \mathrm{PR}\\
&\mathrm{EG} = \mathrm{Energy(kWh)}\\
&\mathrm{Area} = \text{Total solar panel Area}\left(\mathrm{m}^2\right)\\
&n = \text{solar panel yield}\\
&\mathrm{Av} = \text{Annual average solar radiation on tilted panels}\\
&\mathrm{PR} = \text{Performance ratio, coefficient for losses}
\end{aligned}
$$

Fusion AI

AI = LSTM + Recurrent neural network (RNN)

Step 1: Computer the flow of the charging station in terms of storage capacity

Step 2: Determine each storage of charging time slot basis.

Step 3: Define the capacity of the storage specified the time slot

Step 4: Recalculate the illustration's exciting flow by the situation of the highervolume boundary charging stations.

The algorithm has reached its decision.

Because the information on the network link is fully understood, can the problems described above solve optimization of LSTM and RNN Only? Sensors, on the other hand, are prone to malfunctions in the existing IIoT [44] energy network. Loss occurs throughout the transmission procedure as well. As a result, when network information is lacking, it is more feasible to optimize storage allocation algorithms and to route policies together. Another example is that we must plan ahead of time for storage allocation and routing. The only solution for all-electric devicesin the IIoT is to provide RES data. Some future Industry 4.0 and 5.0 bus routing maps for IIoT [45] can be appropriately anticipated from the company's database of Industry 4.0, but buses cannot follow the schedule due to the variety of industry working. This problem cannot be solved in this regard since future network information cannot be obtained. These two types of practical cases are more difficult for existing algorithms. We must take other processes to improve our proposed algorithms to meet our initial goals of maximizing overall energy flow from power generating units to destinations. We will focus our attention on a program controller unaware of industry workingflows in the following sections. Instead, we suppose that the controller calculates industryquantities based on past data.

Furthermore, some of the core nodes can make distributive observations and judgments. We can employ predictive algorithms in some significant schedule controllers to learn industry sensorworking and take historical information. By analyzing these characteristics, the IIoT industry RES pattern [46] is eventually established, and the Industry working may then be projected. Earlier estimation approaches, such as random forest, decision tree, Support vector regression, were all based on IIoT industry RES flow assumptions with unique characteristics. This research provides LSTM and RNN as our training model and compares it to traditional approaches. Because LSTM and RNN can remember historical information while forgetting certain useless historical information in IIoT industry RES forecasting, it is our key prediction. The LSTM and RNN model design adapt to different IIoT industry RES flow features or uniform resource flow prediction. In comparison to the previous IIoT industry RES, we discovered that LSTM and RNN could recognize changing patterns more precisely. we will give a brief introduction for LSTM and RNN model as the background knowledge.

Fig. 6.4: The input layer, LSTM Layer, layer, and output layer are all fully coupled in the LSTM NN topology. For every forecast iteration, y_t is the new expected column sequence value $\{r^1{}_t, r^2{}_t, \ldots, x_t\}$.

Joint optimization of LSTM and RNN prediction The purpose of incomplete data is to increase the network's energy transmission efficiency [47]. Calculating the stochastic network dynamic storage allocation in advance for each charging station is a critical aspect of the solution. To accomplish this, we create an optimizer whose structure. We first take the

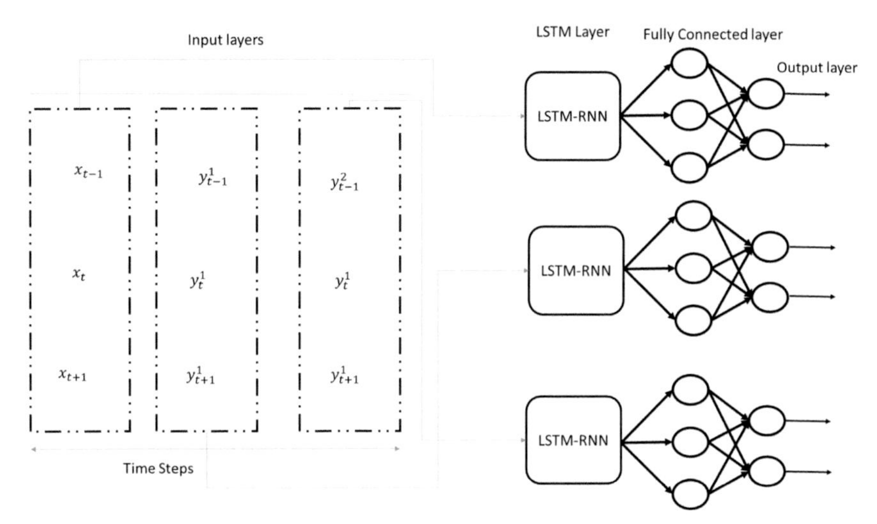

Figure 6.4 Fusion model (long short-term memory-recurrent neural network).

past forecast data as training samples and the historical data of its nearby connections, train the LSTM and RNN model, and then use the learned model to complete and anticipate the IIoT network's missing data. In order to solve the routing and storage optimizing joint problem, this approach simplifies the situation if all of the information is available. After acquiring the dynamic storage allocation and routing system through joint optimization, we apply this result to the genuine IIoT network with missing data. Depicts the method of determining dynamic storage allocation for each charging station. Within the incomplete network, IIoT working flows are separated [48] into different segments. Each section includes the energy resource we wish to predict as well as its nearby energy resource [49]. For each part, we generate an LSTM model. On each section. The missing traffic flow is enabled by using regression to train the LSTM RNN. The network can be reached with complete information by combining a genuine, unfilled network with predicted sensor energy resource [50]. Finally, each charging station has the ability to deploy storage allocations that vary over time. It is compared to the desired result and sent to the LSTM model as feedback. The practice is repeated until the performance improvement is no longer a barrier and the learning notion is strengthened. Understanding the dynamic storage allocation in equates to knowing the storage connection capacity so that the maximum amount of energy can be created by utilizing the maximum flow algorithm from the power station.

6.6 Results analysis

We evaluate different AI algorithms on different IIoT based datasets. Dissimilar assessment metrics were used to examine the goodness of the AI-based model, such as mean absolute error, mean absolute percent error, mean squared error, and root mean squared logarithmic error. We similarly selected the state-of-the-art models for the assessment through the proposed Fusion AI-based model.

6.6.1 Mean absolute error

The mean absolute error (MAE) characterizes the alteration among the original and predictable values and is mined as the dataset's total alteration mean.

$$\mathrm{MAE} = \frac{1}{n}\sum_{i=1}^{n}\left|Y_i - \hat{Y}_i\right|$$

6.6.2 Mean squared error

The mean squared error (MSE) is the alteration between the original value and the predictable value. It is mined by forming the mean formed error of the dataset.

$$\mathrm{MSE} = \frac{1}{n}\sum_{i=1}^{n}\left(Y_i - \hat{Y}_i\right)^2$$

6.6.3 Root mean squared logarithmic error

The root mean squared logarithmic error (RMSLE).

$$\mathrm{RMSLE} = \sqrt{\frac{1}{n}\sum_{i=1}^{n}(\log(\hat{y}_i + 1) - \log(y_i + 1))^2}$$

6.6.4 Mean absolute percent error

The mean absolute percent error (MAPE) is theamount of the accuracy of a prediction. It measures the size of the error (Fig. 6.5; Table 6.1).

$$\mathrm{MAPE} = \frac{\sum \frac{|A-F|}{A} \times 100}{N}$$

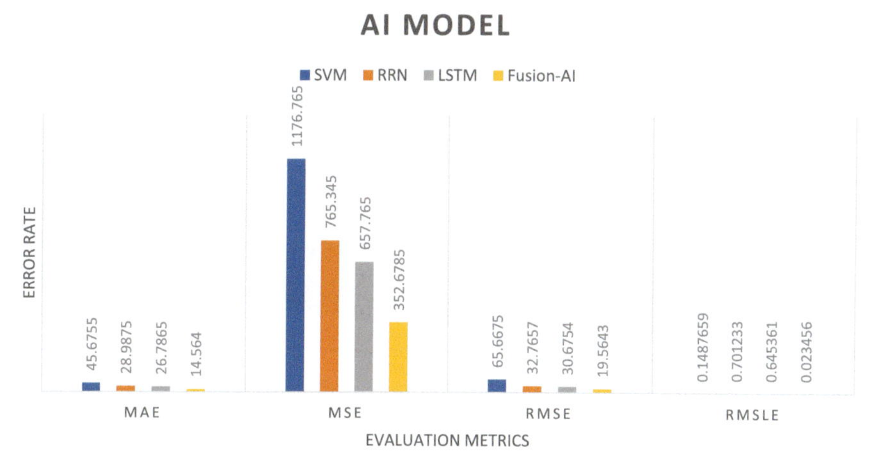

Figure 6.5 Comparative analysis in term of accuracy.

Table 6.1 Evaluation metrics.

Model name	MAE	MSE	Root mean square error	RMSLE
Support Vector Machines (SVM)	45.6755	1176.765	65.6675	0.1487659
Recurrent Neural Network (RNN)	28.9875	765.345	32.7657	0.701233
Long Short-term Memory (LSTM)	26.7865	657.765	30.6754	0.645361
Fusion Artificial Intelligence	14.564	352.6785	19.5643	0.023456

6.7 Conclusion

The IIoT Model Using Fusion-AI of RES has played a significant role in transformingRenewable Energy in this research. IoT digitization improves the efficiency, reliability, resilience, security, and sustainability of electricity grids by increasing accountability forreducing energy waste, saving money, and improving efficiency, dependability, resilience, security, and sustainability. A detailed evaluation of the technical parameters of IoT sensors for the intelligent IIoT scenario was also presented in this research. The customer's value in the energy context comprises safety and reliability and crucial elements such as interoperability and resilience. Power electronics, which plays an essential role in power flow and stability, is one of the two key technology considerations.

In contrast, software, electronics, and embedded systems are primarily used for automation. The systems' major features include customer experience, uncertainty, and the growth of autonomy. The IoT has significantly increased the utilization of renewable energy sources. Power companies to provide a steady supply of electricity to their citizens now use renewable energy. Solar and wind energy usage have already surged because of the IoT. Its use in geothermal, biogas, and hydropower plants should be investigated. Renewable energy is, without a doubt, the way of the future. They will gradually, but unmistakably, meet our rising electrical need.

Reference

[1] G. Yang, J. Cao, H. Hua, Z. Zhou, Deep learning-based distributed optimal control for wide area energy internet, in: 2018 IEEE International Conference on Energy Internet (ICEI), 2018, pp. 292–297. Available from https://doi.org/10.1109/ICEI.2018.00060.

[2] H.K. Huneria, P. Yadav, R.N. Shaw, D. Saravanan, A. Ghosh, AI and IOT-based model for photovoltaic power generation, in: S. Mekhilef, M. Favorskaya, R.K. Pandey, R.N. Shaw (Eds.), Innovations in Electrical and Electronic Engineering. Lecture Notes in Electrical Engineering, vol. 756, Springer, Singapore, 2021. Available from: https://doi.org/10.1007/978-981-16-0749-3_55.

[3] E.C. Akbaba, E. Yüce, B.G. Akinoglu, Deep learning algorithm applied to daily solar irradiation estimations, in: 2018 6th International Renewable and Sustainable Energy Conference (IRSEC), 2018, pp. 1–4. Available from https://doi.org/10.1109/IRSEC.2018.8702963.

[4] Y. Belkhier, A. Achour, R.N. Shaw, W. Sahraoui, A. Ghosh, Adaptive linear feedback energy-based backstepping and PID control strategy for PMSG driven by a grid-connected wind turbine, in: S. Mekhilef, M. Favorskaya, R.K. Pandey, R.N. Shaw (Eds.), Innovations in Electrical and Electronic Engineering. Lecture Notes in Electrical Engineering, vol. 756, Springer, Singapore, 2021. Available from: https://doi.org/10.1007/978-981-16-0749-3_13.

[5] P. Singh, S. Bhardwaj, S. Dixit, R.N. Shaw, A. Ghosh, Development of prediction models to determine compressive strength and workability of sustainable concrete with ANN, in: S. Mekhilef, M. Favorskaya, R.K. Pandey, R.N. Shaw (Eds.), Innovations in Electrical and Electronic Engineering. Lecture Notes in Electrical Engineering, vol. 756, Springer, Singapore, 2021. Available from: https://doi.org/10.1007/978-981-16-0749-3_59.

[6] A.S. Rajawat, S. Jain, K. Barhanpurkar, Fusion protocol for improving coverage and connectivity WSNs, IET Wireless Sens. Syst. 11 (4) (2021) 161–168. Available from: https://doi.org/10.1049/wss2.12018.

[7] A. Singh Rajawat, S. Jain, Fusion deep learning based on back propagation neural network for personalization, in: 2nd International Conference on Data, Engineering and Applications (IDEA), Bhopal, India, 2020, pp. 1–7. Available from https://doi.org/10.1109/IDEA49133.2020.9170693.

[8] R.N. Shaw, P. Walde, A. Ghosh. IOT based MPPT for performance improvement of solar PV arrays operating under partial shade dispersion, in: 2020 IEEE 9th Power India International Conference (PIICON) held at Deenbandhu Chhotu Ram University of Science and Technology, Sonepat, India, February 28–March 1, 2020.
[9] G. Kapoor, V.K. Mishra, R.N. Shaw, A. Ghosh, Fault detection in power transmission system using reverse biorthogonal wavelet, in: S. Mekhilef, M. Favorskaya, R.K. Pandey, R.N. Shaw (Eds.), Innovations in Electrical and Electronic Engineering. Lecture Notes in Electrical Engineering, vol. 756, Springer, Singapore, 2021. Available from: https://doi.org/10.1007/978-981-16-0749-3_28.
[10] V. Puri, et al., A hybrid artificial intelligence and internet of things model for generation of renewable resource of energy, IEEE Access. 7 (2019) 111181–111191. Available from: https://doi.org/10.1109/ACCESS.2019.2934228.
[11] G. Bedi, G.K. Venayagamoorthy, R. Singh, R. Brooks, K.-C. Wang, Review of internet of things (IoT) in electric power and energy systems, IEEE Internet Things J. 5 (2) (2018) 847–870. Available from: https://doi.org/10.1109/JIOT.2018.2802704.
[12] H.H.R. Sherazi, L.A. Grieco, M.A. Imran, G. Boggia, Energy-efficient LoRaWAN for industry 4.0 applications, IEEE Trans. Industr. Inform. 17 (2) (2021) 891–902. Available from: https://doi.org/10.1109/TII.2020.2984549.
[13] J. Clairand, M. Briceño-León, G. Escrivá-Escrivá, A.M. Pantaleo, Review of energy efficiency technologies in the food industry: trends, barriers, and opportunities, IEEE Access. 8 (2020) 48015–48029. Available from: https://doi.org/10.1109/ACCESS.2020.2979077.
[14] Ö. Tuttokmaği, A. Kaygusuz, Smart grids and industry 4.0, in: 2018 International Conference on Artificial Intelligence and Data Processing (IDAP), 2018, pp. 1–6. Available from https://doi.org/10.1109/IDAP.2018.8620887.
[15] G. Kapoor, P. Walde, R.N. Shaw, A. Ghosh, HWT-DCDI-based approach for fault identification in six-phase power transmission network, in: S. Mekhilef, M. Favorskaya, R.K. Pandey, R.N. Shaw (Eds.), Innovations in Electrical and Electronic Engineering. Lecture Notes in Electrical Engineering, vol. 756, Springer, Singapore, 2021. Available from: https://doi.org/10.1007/978-981-16-0749-3_29.
[16] M. Kumar, V.M. Shenbagaraman, A. Ghosh, Predictive data analysis for energy management of a smart factory leading to sustainability, in: M.N. Favorskaya, S. Mekhilef, R.K. Pandey, N. Singh (Eds.), Innovations in Electrical and Electronic Engineering. Lecture Notes in Electrical Engineering, vol. 661, 2021. Available from: https://doi.org/10.1007/978-981-15-4692-1_58.
[17] H. Sun, M. Yin, W. Wei, et al., MEMS based energy harvesting for the internet of things: a survey, Microsyst. Technol. 24 (2018) 2853–2869. Available from: https://doi.org/10.1007/s00542-018-3763-z.
[18] A.S. Rajawat, A.R. Upadhyay, Web personalization model using modified S3VM algorithm for developing recommendation process, in: 2nd International Conference on Data, Engineering and Applications (IDEA), Bhopal, India, 2020, pp. 1–6. Available from https://doi.org/10.1109/IDEA49133.2020.9170701.
[19] R.N. Shaw, P. Walde, A. Ghosh, Review and analysis of photovoltaic arrays with differentconfiguration system in partial shadowing condition, Int. J. Adv. Sci. Technol. 29 (9s) (2020) 2945–2956.
[20] S.H. Alsamhi, O. Ma, M.S. Ansari, et al., Greening internet of things for greener and smarter cities: a survey and future prospects, Telecommun. Syst. 72 (2019) 609–632. Available from: https://doi.org/10.1007/s11235-019-00597-1.
[21] T. Song, J. Cai, T. Chahine, et al., Towards smart cities by internet of things (IoT)—a silent revolution in China, J. Knowl. Econ. 12 (2021) 1–17. Available from: https://doi.org/10.1007/s13132-017-0493-x.

[22] R.N. Shaw, P. Walde, A. Ghosh, Enhancement of power and performance of 9x4 PV Arrays by a novel arrangement with shade dispersion, Test. Eng. Manage. 82 (2020) 13136–13146.
[23] R.N. Shaw, P. Walde, A. Ghosh, Effects of solar irradiance on load sharing of integrated photovoltaic system with IEEE standard bus network, Int. J. Eng. Adv. Technol. 9 (1) (2019) 424–429.
[24] C. Tomazzoli, S. Scannapieco, M. Cristani, Internet of Things and artificial intelligence enable energy efficiency, J. Ambient. Intell. Hum. Comput. (2020). Available from: https://doi.org/10.1007/s12652-020-02151-3.
[25] A. Khanna, S. Kaur, Internet of things (IoT), applications and challenges: a comprehensive review, Wirel. Pers. Commun. 114 (2020) 1687–1762. Available from: https://doi.org/10.1007/s11277-020-07446-4.
[26] R.N. Shaw, P. Walde, A. Ghosh, A new model to enhance the power and performances of 4x4 PV arrays with puzzle shade dispersion, Int. J. Innov. Technol. Explor. Eng. 8 (12) (2019) 1–10.
[27] U. Paukstadt, J. Becker, Uncovering the business value of the internet of things in the energy domain – a review of smart energy business models, Electron. Mark. 31 (2021) 51–66. Available from: https://doi.org/10.1007/s12525-019-00381-8.
[28] C. Bordin, S. Mishra, A. Safari, et al., Educating the energy informatics specialist: opportunities and challenges in light of research and industrial trends, SN Appl. Sci. 3 (2021) 674. Available from: https://doi.org/10.1007/s42452-021-04610-8.
[29] S. Muralidharan, A. Roy, N. Saxena, An exhaustive review on internet of things from Korea's perspective, Wireless Pers. Commun. 90 (2016) 1463–1486. Available from: https://doi.org/10.1007/s11277-016-3404-8.
[30] A. Mukherjee, P. Mukherjee, D. De, et al., iGridEdgeDrone: hybrid mobility aware intelligent load forecasting by edge enabled internet of drone things for smart grid networks, Int. J. Parallel Prog. 49 (2021) 285–325. Available from: https://doi.org/10.1007/s10766-020-00675-x.
[31] H. Chen, X. Wang, Z. Li, et al., Distributed sensing and cooperative estimation/detection of ubiquitous power internet of things, Prot. Control. Mod. Power Syst. 4 (13) (2019). Available from: https://doi.org/10.1186/s41601-019-0128-2.
[32] T. Xue, X. Liu, Y. Zeng, Y. Zhang, Resilient event-triggered controller synthesis of load frequency control for multi-area power systems under periodic DoS jamming attacks, in: 2019 Chinese Control Conference (CCC), 2019, pp. 5433–5438. Available from https://doi.org/10.23919/ChiCC.2019.8866688.
[33] K. Majumder, K. Chakrabarti, R.N. Shaw, A. Ghosh, Genetic algorithm-based two-tiered load balancing scheme for cloud data centers, in: J.C. Bansal, L.C.C. Fung, M. Simic, A. Ghosh (Eds.), Advances in Applications of Data-Driven Computing. Advances in Intelligent Systems and Computing, vol. 1319, Springer, Singapore, 2021. Available from: https://doi.org/10.1007/978-981-33-6919-1_1.
[34] T.S.L.V. Ayyarao, I.R. Kiran, A two-stage kalman filter for cyber-attack detection in automatic generation control system, J. Modern Power Syst. Clean Energy (2021). Available from: https://doi.org/10.35833/MPCE.2019.000119.
[35] P.I. Radoglou-Grammatikis, P.G. Sarigiannidis, Securing the smart grid: a comprehensive compilation of intrusion detection and prevention systems, IEEE Access. 7 (2019) 46595–46620. Available from: https://doi.org/10.1109/ACCESS.2019.2909807.
[36] S. Bodapati, H. Bandarupally, R.N. Shaw, A. Ghosh, Comparison and analysis of RNN-LSTMs and CNNs for social reviews classification, in: J.C. Bansal, L.C.C. Fung, M. Simic, A. Ghosh (Eds.), Advances in Applications of Data-Driven Computing. Advances in Intelligent Systems and Computing, vol. 1319, Springer, Singapore, 2021. Available from: https://doi.org/10.1007/978-981-33-6919-1_4.

[37] C. Choi, J. Jeong, I. Lee, W. Park, LoRa based renewable energy monitoring system with open IoT platform, in: 2018 International Conference on Electronics, Information, and Communication (ICEIC), 2018, pp. 1–2. Available from https://doi.org/10.23919/ELINFOCOM.2018.8330550.
[38] C. Shin, S. Lee, J. Kim, H. Nam, Y.K. Jeong, A study on the implementation of economic zero energy building according to Korea's renewable energy support policies and energy consumption patterns, in: 2018 International Conference on Information and Communication Technology Convergence (ICTC), 2018, pp. 1305–1309. Available from https://doi.org/10.1109/ICTC.2018.8539571.
[39] S. NurAltun, M. Dörterler, I. AlperDogru, Fuzzy logic based lighting system supported with IoT for renewable energy resources, in: 2018 Innovations in Intelligent Systems and Applications Conference (ASYU), 2018, pp. 1–4. Available from https://doi.org/10.1109/ASYU.2018.8554026.
[40] E. López, J. Monteiro, P. Carrasco, J. Sáenz, N. Pinto, G. Blázquez, Development, implementation and evaluation of a wireless sensor network and a web-based platform for the monitoring and management of a microgrid with renewable energy sources, in: 2019 International Conference on Smart Energy Systems and Technologies (SEST), 2019, pp. 1–6. Available from https://doi.org/10.1109/SEST.2019.8849016.
[41] D.K. Aagri, A. Bisht, Export and import of renewable energy by hybrid microgrid via IoT, in: 2018 3rd International Conference on Internet of Things: Smart Innovation and Usages (IoT-SIU), 2018, pp. 1–4. Available from https://doi.org/10.1109/IoT-SIU.2018.8519873.
[42] C. Nayanatara, S. Divya, E.K. Mahalakshmi, Micro-grid management strategy with the integration of renewable energy using IoT, in: 2018 International Conference on Computation of Power, Energy, Information and Communication (ICCPEIC), 2018, pp. 160–165. Available from https://doi.org/10.1109/ICCPEIC.2018.8525205.
[43] Y. Guan, W. Feng, Y. Wu, J.C. Vasquez, J.M. Guerrero, An IoT platform-based multi-objective energy management system for residential microgrids, in: 2020 IEEE 9th International Power Electronics and Motion Control Conference (IPEMC2020-ECCE Asia), 2020, pp. 3107–3112. Available from https://doi.org/10.1109/IPEMC-ECCEAsia48364.2020.9368001.
[44] M. Moness, A.M. Moustafa, A survey of cyber-physical advances and challenges of wind energy conversion systems: prospects for internet of energy, IEEE Internet Things J. 3 (2) (2016) 134–145. Available from: https://doi.org/10.1109/JIOT.2015.2478381.
[45] G. Muthuselvi, B. Saravanan, The promise of DSM in smart grid using home energy management system with renewable integration, in: 2017 Innovations in Power and Advanced Computing Technologies (i-PACT), 2017, pp. 1–4. Available from https://doi.org/10.1109/IPACT.2017.8245102.
[46] Z. Hameed, F. Ahmad, S.U. Rehman, Z. Ghafoor, IoT based communication technologies to integrate and maximize the efficiency of renewable energy resources with smart grid, in: 2020 International Conference on Computing and Information Technology (ICCIT-1441), 2020, pp. 1–5. Available from https://doi.org/10.1109/ICCIT-144147971.2020.9213730.
[47] L. Liu, H. Sun, P. Gao, N. Zheng, T. Li, REcache: efficient sustainable energy management circuits and policies for computing systems, in: 2019 IEEE International Symposium on Circuits and Systems (ISCAS), 2019, pp. 1–5. Available from https://doi.org/10.1109/ISCAS.2019.8702515.
[48] U. Hijawi, A. Gastli, R. Hamila, O. Ellabban, D. Unal, Qatar green schools initiative: energy management system with cost-efficient and lightweight networked IoT,

in: 2020 IEEE International Conference on Informatics, IoT, and Enabling Technologies (ICIoT), 2020, pp. 415–421. Available from https://doi.org/10.1109/ICIoT48696.2020.9089443.
[49] M. Kumar, A.F. Minai, A.A. Khan, S. Kumar, IoT based energy management system for smart grid, in: 2020 International Conference on Advances in Computing, Communication & Materials (ICACCM), 2020, pp. 121–125. Available from https://doi.org/10.1109/ICACCM50413.2020.9213061.
[50] A.R. Al-Ali, Internet of things role in the renewable energy resources, Energy Procedia 100 (2016) 34–38. Available from: https://doi.org/10.1016/j.egypro.2016.10.144.

CHAPTER SEVEN

Centralized intelligent fault localization approach for renewable energy-based islanded microgrid systems

Ahteshamul Haque, V. S. Bharath Kurukuru, Mohammed Ali Khan, Azra Malik and Faizah Fayaz
Advance Power Electronics Research Lab, Department of Electrical Engineering, Jamia Millia Islamia (A Central University), New Delhi, India

7.1 Introduction

Growing electricity demand together with the need for reduction of environmental pollution caused by conventional electricity production have instigated the research towards microgrids. Further, islanded microgrids with renewable sources is considered as an effective solution to these concerns. However, the technical difficulties with these small-scale inverter-based islanded microgrids create new challenges to achieve better stability, quality, and reliability of supply [1]. This chapter explores such issues in islanded microgrids and proposes near real-time disturbance detection and protective solutions for their stable operation.

The stability issue is an area of concern during the islanded microgrid operation [2–4]. The microgrid operating in islanded mode should be smart enough to control the voltage, system frequency and achieve power balance. As the islanded mode of operation for the distributed generation depends upon the power electronic converter, hence there is not much of inertia for reserving the kinetic energy and enduring sudden changes [5]. Because of the low physical inertia, the dynamic response of islanding microgrid is faster with respect to conventional rotating machine which led to reduce the susceptibility to oscillation for network disturbances [6,7]. Hence, the rate of change in frequency (RoCoF) act after the disturbance and before the control action is initialize, and the resultant is

Applications of AI and IOT in Renewable Energy.
DOI: https://doi.org/10.1016/B978-0-323-91699-8.00007-3

high in case of islanded microgrid [8]. Hence the boundary limits are easily achieved for the system frequency [9]. During the short circuit fault, the power electronics converters are not capable enough to handle high fault current for a long duration [10]. Hence the time for identification of the disturbance is limited to few voltage cycles [11]. The process is time critical and distribution detection techniques is need to be highly time intensive which is necessary for islanded microgrid operation.

Major disturbances such as short circuit faults and step changes in loads/generations in islanded microgrids heavily affect the system stability, quality as well as reliability. The behavior of islanded microgrids during these disturbances significantly vary from the conventional system behavior [12]. Importantly, the time available to arrest these disturbances for further actions is very less compared to a conventional system [13,14]. Overcurrent, differential and distance protections are well established and accurate short circuit fault detection techniques in conventional power systems. However, these methods fail to perform in islanded microgrids due to insignificant fault current, less system inertia, quick response of power electronic inverters, and the issue of bidirectional power flow [15–17]. Also, unlike in conventional system, system-wide RoCoF and swing equation cannot be used to estimate the power imbalance in islanded microgrids following the step changes in loads/generations due to their highly resistive distribution lines which create location-specific behavior of initial RoCoF andsystem inertia variation with time [18]. An accurate power imbalance estimation is undeniably needed to decide the load shedding amount to get back the system power balance.

Conventional disturbance detection techniques are model/topology-based post event analysis which are not applicable for islanded microgrids since they react to disturbances much quicker than conventional system [19]. Real-time disturbance detection using data-driven approaches are required for islanded microgrids. This is feasible as the currently available measurement sensors provide enough data to learn the system behavior [20]. Recently, machine learning algorithms together with advanced digital signal processing tools have gained ever more attention in islanded microgrid operation [21].

State-of-the-art machine learning-based disturbance detection methods require enough data to learn the system disturbances which will be consequently used to detect the disturbances in real-time [22]. It needs an informative feature from system frequency, voltage and/or current signal to represent the system behavior. Digital signal processing tools have an ability to capture even the minor transients in a signal either in the time domain or frequency domain.

Islanded microgrids require these data-driven disturbance detection approaches to enable them with time-intensive adaptive features.

The development of algorithms that enable real-time disturbance detection, and classification for islanded AC microgrids are the underlying focus of the work in this chapter. The principal objective of this work is to develop a novel disturbance detection and protective method for islanded AC microgrids by investigating the use of digital signal processing tools and machine learning techniques. Digital signal processing tools have been used to study the transient behavior of system frequency, voltage and current signals in terms of detecting and investigating the system disturbances. Machine learning techniques are used for disturbance detection, classification, and power imbalance calculation. This chapter mainly focuses on short circuit faults and step changes in loads/generations to detect and classify them in real-time for further protection actions. The major contributions of the chapter are:

To examine the behavior of frequency, voltage and current signals following disturbances in islanded microgrids.

To understand and investigate the existing islanded microgrid disturbance detection methods specially to understand their performance in a wide range of disturbance parameters such as disturbance location, disturbance amount, fault impedance, etc.

To develop and validate real-time fault detection and classification techniques and examine the feasibility of the developed methods.

Further sections of the chapter are organized as follows: Section 7.2 provides a brief discussion on the challenges in the islanded microgrid framework during disturbance detection. Section 7.3 identifies the requirements for developing the proposed localization approach. Section 7.4 develops the proposed methodology as a centralized voltage signal-based fault classification approach. The numerical simulation, results, and discussions are made in Section 7.5, and the conclusion is given in Section 7.6.

7.2 Challenges in disturbance detection

7.2.1 Behavior of power electronics converters

The environmentally friendly renewable based DERs such as wind, and solar are rapidly increasing in power systems to decrease greenhouse gas emission from fossil fuels. However, the high penetrations of these DERs create new technical challenges to the power system due to their low inertia, and the behavior of their power electronic inverters. Generally, DERs can be classified as two categories that are (1) DERs connected to the grid directly, and (2) inverter interfaced DERs. Most of the DERs such as solar plants, direct-drive wind turbines, batteries, flywheel energy storage, and micro gas turbines generate DC power which is converted to AC and interfaced to the AC networks with power electronic inverters. These inverters are made up of semiconductor components. They allow bidirectional power flow with insulated-gate bipolar transistors and diodes connected in the reverse direction. The use of metal-oxide-semiconductor field-effect transistors offer better voltage operation, greater power gain with less power loss and, higher allowable junction temperature [23].

However, since the response time of power electronic inverters is much faster than the conventional generations, the inverter interfaced microgrids' transients in frequency, voltage, and current following any disturbance are also quick. This results in high RoCoF immediately after disturbances in renewable-based systems compared to conventional systems [24]. In addition, to that, output impedance and overcurrent capability of power electronic inverters are very small compared to conventional synchronous generators. They do not have the capability to handle high current flow through them. Hence, the protection response time of these inverter-based DERs is much faster which results in disconnection of these inverters when the microgrid is subjected to large disturbances like faults. This is especially true for low impedance faults where the fault current is very high and may cause fatal damage to the inverters if left to ride-through [25,26]. Therefore the characteristics of power electronic inverters employed in islanded microgrids bring additional challenges in disturbance detection and make it as an extreme time-critical process.

7.2.2 Other disturbances and detection challenges

Islanded AC microgrids are highly nonlinear systems which comprise of different devices such as a variety of generating units, active and reactive loads with multiple configurations, distribution lines and controllers [27]. There are lots of possible disturbances in the system originating from any of those devices. The minor disturbances like insignificant fluctuations in loads and generations are described as small-signal stability issues, and major disturbances, like short circuit faults, step changes in load/generations, switching of loads/generations, etc., are described as transient stability issues. The major disturbances can be planned or unplanned. The planned disturbances like switching of generations, load shedding, etc., can be handled with a planned control action. On the other hand, the microgrid system should have an ability to automatically detect and provide a protective solution in real-time for the unplanned disturbances like short circuit faults and step changes in loads/generations. This research mainly focuses on major disturbances' detection such as short circuit faults and step changes in loads/generations and their protective solutions.

7.3 Requirements for classifier development

Conventional disturbance detection techniques are incapable to perform in islanded microgrids due to the differed characteristics of power electronic inverters and less inertia in renewable sources deployed in islanded AC microgrids. Even though the conventional techniques are modified to perform adaptively, the time-varying inertia, and quick response of power electronic inverters do not allow those adaptive methods to perform well. Hence, there is a necessity to research for an appropriate disturbance detection technique that should be capable enough of performing well while coping up with the quick response of power electronic inverters in islanded microgrids. In light of this requirement, the digital signal processing techniques are identified as fast and sensitive in detecting the transients in the signals measured in power systems. They process the measured digital signals to detect the system disturbances. The processed signals are considered for feature extraction where the most informative features are extracted to represent the signal behavior. The extracted features can be used in further analysis such as disturbance detection, classification, and localization where machine learning algorithms come into play.

7.3.1 Feature extraction

Feature extraction from a measured signal is a critical task for automatic decision making of machine learning design and algorithm. Usually in power systems, a real-time moving window is applied on voltage, current and/or frequency signals and the windowed signals are considered for feature extraction. It transforms the high dimensional data to fewer dimensional feature vector which is expected to be informative as well as nonredundant. It can be related to dimensional reduction. While the input data to the machine learning algorithm is too large, or it consists of very much redundant information, the raw data can be transformed into a reduced set of features which is known as feature extraction. The extracted features represent the signal behavior, which are inputs to the machine learning algorithms to develop the classifier or regression model, and the same features extracted from the unseen data will be used to predict the output.

In a power system, the features are commonly extracted from the measured signals such as frequency, voltage, and current. An appropriate feature extraction algorithm needs to be developed in order to extract as much as information from the measured signal. Since digital signal processing techniques have an ability to capture fast transients in power system measurements, they can be used to extract features in power system applications. The features can be extracted in the time domain, and/or frequency domain of the measured signals using different transformation techniques. For example, subband energy from Fourier transform [28], or statistical features from wavelet coefficients at different decomposition level [29], or instantaneous jumps in system voltage using Hilbert Transform (HT) can be extracted as features. Since extracting an informative feature significantly influences the performance of machine learning algorithms, developing an appropriate feature extraction algorithm is of prime importance. Especially in power system applications, since these machine learning algorithms run in real-time, the computational complexity from the feature extraction process should be minimized to avoid delays in protection actions. Eventually, the extracted features are used to develop the classifier or regression models.

In this research, the time domain operation is implemented with the HT due its efficiency in identifying the instantaneous changes and range of the change. The HT $\hat{x}(t)$ of a discrete-time series signal $x(t)$ is defined as the convolution of $h(t)$ with $\frac{1}{t}$ using the Cauchy principal value and it can be expressed as:

$$x(t) = \frac{1}{\pi} P \int_{-\infty}^{+\infty} \frac{x(t)}{t - \tau} d\tau \tag{7.1}$$

where P indicates the Cauchy principal value. This transform is possible for all integrable functions.

7.3.2 Machine learning

Machine learning helps to learn the system behavior from the inputs available and then converts the experience into expertise or knowledge [30]. The input to a learning algorithm is called as training data which represents the experience, and the output is knowledge, which usually takes the form of another computer program that can perform some tasks. These machine learning algorithms provide predictions based on known properties learned from the training. They perform classification, regression, clustering, density estimation, and dimensionality reduction. The main focus of this research is the classification process.

Classification is the process of predicting the class of given input data. A classifier model should be trained in advance with known inputs, later the trained classifier model assigns unseen inputs to one or more classes. This is usually tackled with supervised learning methods. For example, the type of short circuit faults in a power system can be identified by performing machine learning-based classification. There are many machine learning classification algorithms available such as Support Vector Machine (SVM), Artificial Neural Network, Decision Tree, Random Forest, etc. All these machine learning models require an appropriate feature vector as input to perform the classification or regression. Hence feature extraction is an important task in machine learning algorithms [30]. Further, in this research, the SVM is used to train the classifier due to its simplicity in handling the data, fast training and testing capability.

The SVM is a supervised learning algorithm which is associated to perform both classification and regression. The objective of the SVM classification algorithms is to find a hyperplane in an N-dimensional space that clearly classifies the data points. In a k-dimensional feature space, the hyperplane is a subspace of dimension $k-1$. For example, in a 2D feature space, the hyperplane is a line, and in a 3D featurespace, it is a flat plane. If the number of features is more than three, it becomes complicated to visualize but the SVM has an ability to define the hyperplane for any number of features. For example, to separate data points with two classes in 2D plane, many possible hyperplanes could be chosen. However, the objective here is to choose the plane with maximum margin. Maximizing the margin distance gives better reinforcement which increases the confidence level in classification of future data points. These hyperplanes are the decision boundaries which classify the data points into suitable classes. The support vectors are the data points that lie very close to the hyperplane margin. Classifying these support vectors becomes harder compared to other data points. Hence, the decision function is completely specified by the support vectors which are a subset of training samples.

If the data is linear, the hyperplane can be easily identified to separate each class. On the other hand, if the data is nonlinear and inseparable with linear planes, kernel functions map the nonlinear input data to a high-dimensional space to make that as linearly separable. Hence, the selection of suitable kernel function becomes an important task in SVM.

7.4 Centralized fault localization method

The insignificant fault currents in the operation of a DG inverter have made the fault detection a challenging task. Hence, the voltage signal-based fault localization method is considered in this proposed approach for monitoring the islanded microgrids. Through the process of observing the effect of various faults on the measured voltage it is identified that, there an instant change in the phase angle, amplitude, and frequency of the voltage immediately after a fault. This corresponding observation is shown in Fig. 7.1. In light of these instant changes, the proposed approach is motivated at the development of the fault classification approach by analyzing the features and training the classification models. As discussed in the previous section, the HT algorithm is used for feature extraction of the instant changes in a time domain. This approach is capable of differentiating the features in various scenarios ranging from step changes generations and loads, tovarious types of symmetrical and

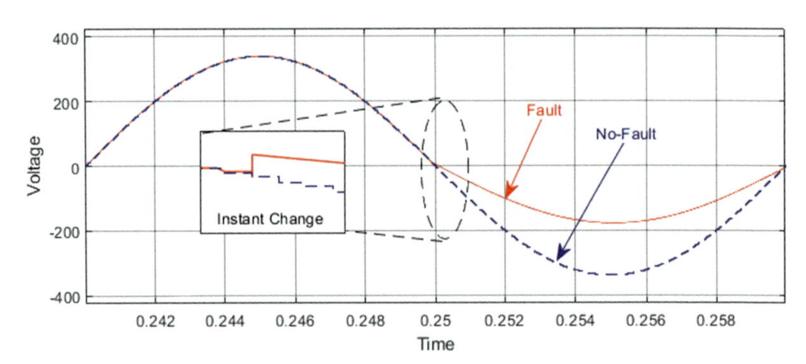

Figure 7.1 Instant change in measured signal during fault condition.

unsymmetrical faults. Further, the SVM technique as discussed in Section 7.3 is used to train the extracted features for efficient and fast localization of the faults and disturbances in an islanded microgrid.

7.4.1 Data gathering

To begin with, a moving window of 20 ms and 50% overlap is considered to identify the changes in a fundamental voltage cycle of the measured 3ϕ per unit voltage signals. The feature extraction process is carried out in this time frame for all the measured voltages. Besides, the extracted features are assisted by the RoCoF feature which is estimated for each cycle as given in Eq. (7.2) to classify the fault and disturbance type in the system.

$$\mathrm{RoCoF} = \frac{f_{\mathrm{start}} - f_{\mathrm{end}}}{\Delta t} \tag{7.2}$$

where, f_{start} is the frequency at the start of the time window, f_{end} is the frequency at the end of the time window, and Δt indicates the window length that is, 20 ms. The process of the feature extraction with the HT algorithm is explained in detail in the further sections.

7.4.1.1 Feature extraction

The HT algorithm analyses the measured voltage cycle in a time domain, and has the ability to estimate the instant changes in phase angle, amplitude, and frequency. These instantaneous changes are highly dependent on the fault impedance and other fault parameters and provide a lot of information about the fault. Given a voltage signal $v(t)$,

$$Z(t) = v(t) + i\hat{v}(t) = \mathrm{Amp}(t)e^{i\mathrm{Ph}(t)}, \tag{7.3}$$

where,

$$\mathrm{Amp}(t) = \sqrt{v^2(t) + \hat{v}^2(t)}, \tag{7.4}$$

$$\mathrm{Ph}(t) = \arctan\left(\frac{\hat{v}(t)}{v(t)}\right). \tag{7.5}$$

The instant change in angular frequency $\mathrm{rad}^{-1}(\omega(t))$ is estimated as

$$\omega(t) = \frac{d\mathrm{Ph}(t)}{dt}, \tag{7.6}$$

And the instant change in frequency Hz($f(t)$) is estimated as

$$f(t) = \frac{f_s}{2\pi}\omega(t). \tag{7.7}$$

Here, f_s indicates the sampling frequency.

The flow diagram of the feature extraction process with the HT algorithm is shown in Fig. 7.2. The feature extraction process is implemented separately with all the three phases in the measured voltage. The instant amplitude, phase angle, and frequency of the measured signal is estimated using (7.4) to (7.7), respectively. They form an $\frac{f_s}{f_0}$ dimensional vector, with f_0 as the fundamental frequency of the system.

Further, the instant change in amplitude (ΔAmp(t)), phase angle (ΔPh(t)) and frequency ($\Delta f(t)$) are estimated as follows:

$$\Delta \text{Amp}(t) = \text{Amp}(t+1) - \text{Amp}(t), \tag{7.8}$$

$$\Delta \text{Ph}(t) = \text{Ph}(t+1) - \text{Ph}(t), \tag{7.9}$$

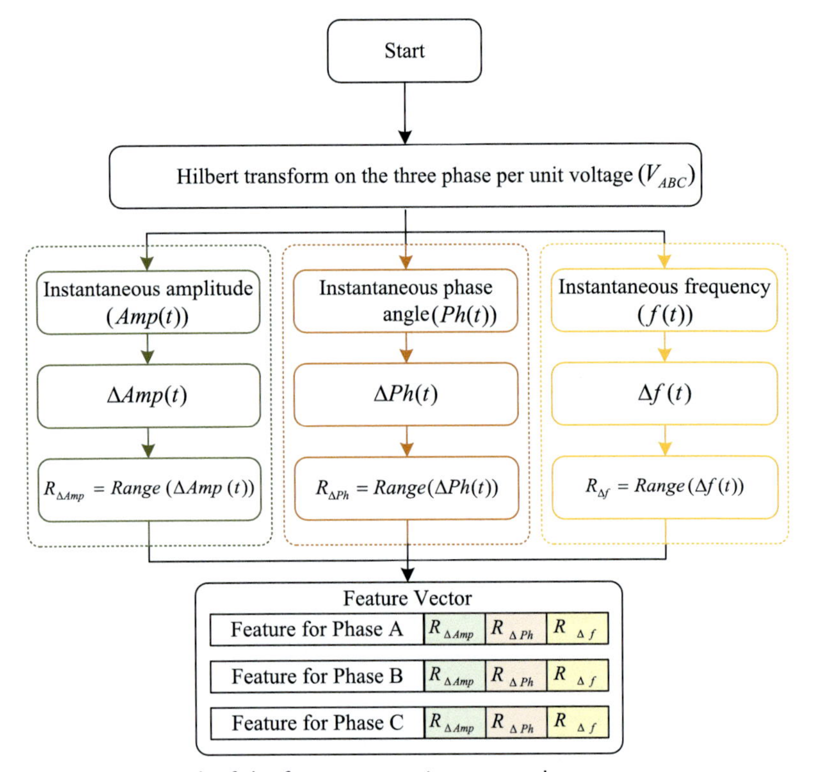

Figure 7.2 Framework of the feature extraction approach.

$$\Delta f(t) = f(t+1) - f(t). \tag{7.10}$$

These features can only be used for identifying the fault in the system, but for fault classification, the range of change is estimated as $R_{\Delta \mathrm{Amp}}$, $R_{\Delta \mathrm{Ph}}$, and $R_{\Delta f}$ as shown in Fig. 7.2. All the extracted features for each phase voltage correspond to form a total nine features in the feature matrix for each fault/disturbance and normal operation.

Further, based on the extracted features for both fault/disturbance and normal operating conditions of a three-phase islanded microgrid system, the fault localization approach is modeled using the machine learning classifiers.

7.4.2 Fault/disturbance detection

In this section, a three-step fault classifier is proposed as shown in Fig. 7.3 to detect faults/disturbances and localize them in an islanded microgrid operation. In the first step of the classification process, the SVM classifier is used to train the nine features extracted for all the three phases to identify whether the microgrid is operating in a normal state of in a fault/disturbance state. This basically involves two classes, class 1 indicates the normal operation of the system, and class 2 indicates all the disturbed operation of the system

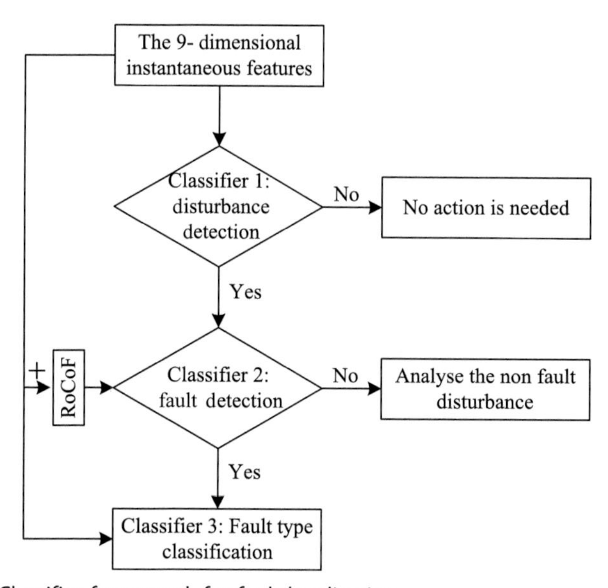

Figure 7.3 Classifier framework for fault localization.

which indicates all types of faults, nonfault transients, and step changes in generations/loads, etc. This step doesn't differentiate different types of faults and hence acts an immediate monitoring response to any change in the operation of the system. Further, the step 2 of the fault classifier involves another SVM classifier which process the identified fault/disturbances in the step 1 classifier to localize the abnormality to fault or disturbance in the system. During this stage of fault classification, the initial nine features of the classifier input are accompanied by the RoCoF feature. This feature helps the classifier in differentiating between the fault and disturbance and is padded as the 10th feature for the fault/disturbance classification. Here, if the detected anomaly is a disturbance a nontransient fault, then it will be analyzed separately which is the future aspect of this research. Further, if the detected anomaly is a fault, then the detected signal is forwarded to step 3, that is, the localization process. In this process, a new classifier is trained with the feature data which is initially used at the input of the classifier 1, and the output from the classifier 2. In this step of classification process, the SVM classifier is trained as discussed in Section 7.3 to localize the fault type to symmetrical, or unsymmetrical, along with the information of the fault location in the system. Here the symmetrical faults involve $A-B-C-G$, and $A-B-C$, which are labeled as class 1. Similarly. the unsymmetrical faults involve $A-B-G, A-C-G, B-C-G, A-B, A-C, B-C, A-G,$ $B-G, C-G$. and other faults are labeled as class 2 to class 11, respectively.

7.5 Numerical simulations

To analyze the operation of the proposed localization approach in a microgrid network, a 3-bus islanded microgrid framework is simulated as shown in Fig. 7.4. The system adapts the characteristics of an AC microgrid model as per the IEC 61850–7–420 regulation [31]. The modeled framework is a three-phase balanced system with a PV array, diesel generator, distribution lines, and connected loads. The PV array is modeled with the SunPower SPR-305-WHT PV module available in the MATLAB/Simulink model library. The power delivered by the PV array at standard test conditions is 300 kW. Further, the diesel generator connected in the system is a $3-\phi$ synchronous machine rated at 200 kVA and operates in a dq reference frame. An automatic voltage regulator controls the rotor current in the system which in turn controls the terminal voltage of the generator [31]. In addition to the above, two

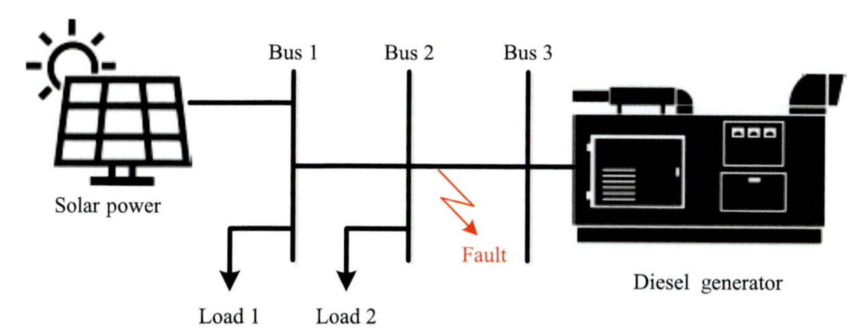

Figure 7.4 Islanded microgrid network.

$3-\phi, 300$ kW, 40 kVAR balanced loads are modeled with the system. Further, the distribution lines are standard PI line models of length 200 m in line 1–100 m in line 2. The zero-sequence resistance, inductance, and capacitance of distribution lines are, $R_0 = 0.1153\Omega/\text{km}$, $L_0 = 1.05\times10^{-3}$ H/km, and $C_0 = 11.33\times10^{-9}$ F/km, respectively. Similarly, the positive sequence resistance, inductance, and capacitance of distribution lines are, R_1=$0.413\Omega/\text{km}$, $L_1 = 3.32\times10^{-3}$H/km, and $C_1 = 5.11\times10^{-9}$ F/km, respectively.

7.5.1 Data collection

The data collection process is implemented by operating the system under variable irradiance condition $(0-1000)\text{W/m}^2$ and variable load condition $(150-300)$ kW. In each condition, multiple disturbances and fault events are created, and their corresponding voltage is measured at the swing bus. Further, the measured voltages are processed at 20 kHz sampling frequency. The details of different fault conditions and the corresponding data samples are tabulated as shown in Table 7.1.

As the low impedance faults doesn't have any impact on the renewable energy-based DG system, the high impedance faults are considered with a fault impedance up to 100Ω in the system. Further, to emulate the near-real-time behavior of the measured data, an additive white gaussian noise of 30 decibels is injected in to the measured data of all the conditions. A sample of voltage signal measured for different abnormalities and operating conditions is shown in Fig. 7.5.

The measured data is subjected to fault classification as discussed in Section 7.4. Three different features are extracted for all the three phases

Table 7.1 Parameters utilized for data gathering and fault classification.

Parameter	Description
Faults considered	Symmetrical faults: $\mathbf{A-B-C-G}$, and $\mathbf{A-B-C}$ Unsymmetrical faults: $\mathbf{A-B-G, A-C-G, B-C-G, A-B, A-C, B-C, A-G, B-G, C-G}$, and others
Fault resistance	$0.1\Omega - 100\Omega$
Load variation	$\pm 20\%$
Irradiance variation	$\pm 400 W/m^2$
Fault location	For each line $5\% - 95\%$ length

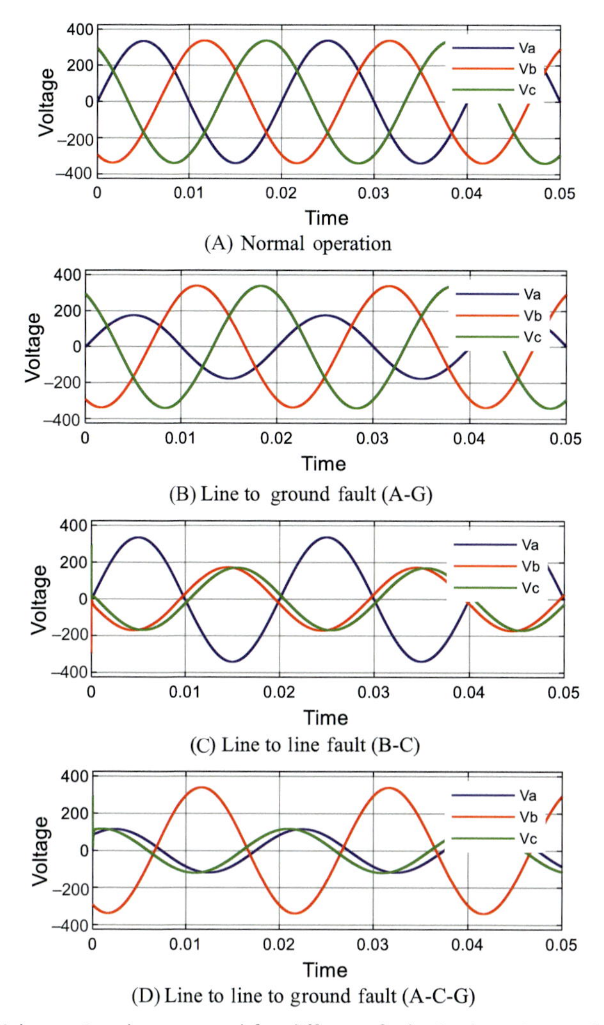

(A) Normal operation

(B) Line to ground fault (A-G)

(C) Line to line fault (B-C)

(D) Line to line to ground fault (A-C-G)

Figure 7.5 Voltage signals measured for different faults in the microgrid framework.

which form up to nine features for each fault case. The details of samples used for training all the three classifiers is given in Table 7.2.

7.5.2 Results and discussion

The SVM classifier is used to train all the three classifiers. As the data collected is nonlinear, the SVM implements adapts a kernel approach to train the classifier. In this approach, the fine gaussian kernel is used map the nonlinear data into the high dimensional feature space to train the

Table 7.2 Parameters for classifier training.

Classifier type	Parameter	Data
Classifier 1 data: SVM classifier	Number of classes	2
	Number of samples	20,000
	Number of features	9
	Class types	Normal operation, and abnormal operation
Classifier 2 data: SVM classifier	Number of classes	2
	Number of samples	10,000
	Number of features	10
	Class types	Disturbance, and fault
Classifier 3 data: SVM classifier	Number of classes	11
	Number of samples	5000
	Number of features	9
	Class types	1 symmetrical fault class, 10 unsymmetrical fault class

classifier. The testing and training results corresponding to the training of the classifier in all the three steps of the proposed approach are shown in Fig. 7.6.

From the results in Fig. 7.6, it is identified that the adapted SVM classifier efficiently trains the feature matrix for classifying and localizing the fault. The confusion matrix in Fig. 7.6A shows the truly and falsely classified samples while training the classifier with the features of normal operation and abnormal operations. The true positive rate and false negative rate of the trained classifier is shown in Fig. 7.6B. From the results it is identified that 94.85 samples are truly classified for normal operation, and 95.3% samples are truly classified for abnormal condition. All the remaining samples in this classification step are misclassified or falsely classified. The overall accuracy of the classification process in step 1 is 95%. The fault detection time with the trained classifier is 0.25 ms. Similarly, the confusion matrix in Fig. 7.6C shows the truly and falsely classified samples

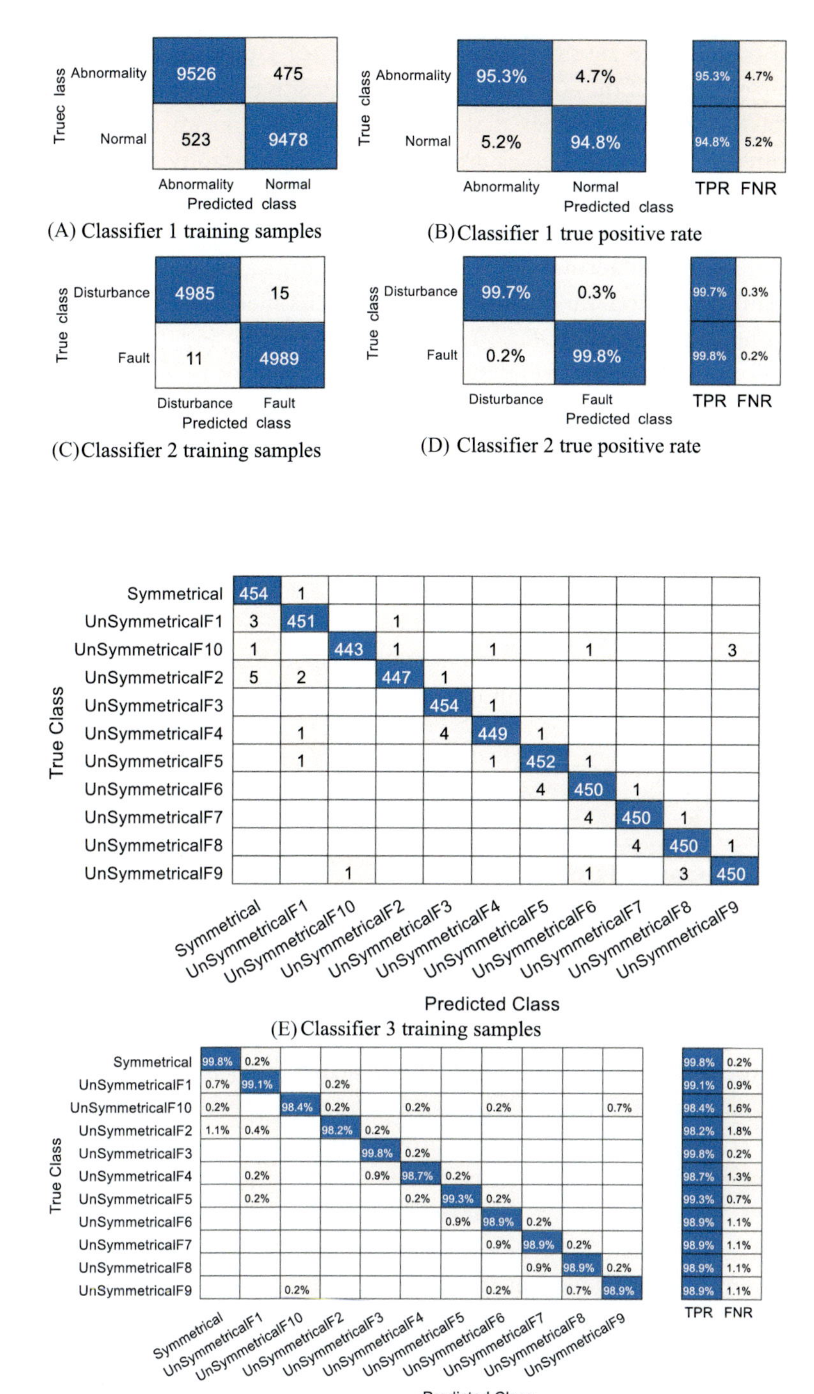

Figure 7.6 Training results of the support vector machine classifier for the three steps of the proposed approach.

while training the classifier with the features of faulty operation and disturbance in operation conditions. The true positive rate and false negative rate of the trained classifier is shown in Fig. 7.6D. From the results it is identified that 99.8% samples are truly classified for faulty operation, and 99.7% samples are truly classified for disturbance condition. All the remaining samples in this classification step are misclassified or falsely classified. The overall accuracy of the classification process in step 2 is 99.8%. The fault detection time with the trained classifier is 0.25 ms. Further, the confusion matrix in Fig. 7.6E shows the truly and falsely classified samples while training the classifier with the features of faulty operation and disturbance in operation conditions. The true positive rate and false negative rate of the trained classifier is shown in Fig. 7.6F. From the results it is identified that large misclassification occurs for data corresponding to unsymmetrical faults 2 and 10 which is greater than 1.5% when compared with the misclassification in other classes. The overall accuracy of the classification process in step 3 is 99%. The fault detection time with the trained classifier is 0.4 ms.

From the above results it is clearly identified that the trained classifier can efficiently classify the faults and localize them with in 0.9 ms.

7.6 Conclusion

This chapter developed an anomaly localization approach to provide a protective solution for islanded AC microgrids. This work has addressed misclassification of fault and disturbances in an islanded microgrid network by formulating a centralized voltage signal-based fault localization method. The potential importance behind this research is to introduce near-real-time, adaptive, anomaly localization techniques under noisy and varying system conditions, by minimizing unwanted tripping and load shedding to improve the system operation capability. Initially, the characteristics of an islanded AC microgrid are discussed and compared with the conventional grid. It is identified that, in case of short circuit faults, the power electronic inverters employed in islanded microgrids are not capable of carrying high fault currents for a long time. Due to the quick response of power electronic inverters in islanded microgrids, fault transients are much faster than the conventional grid. To this end, the time available to arrest the fault is very less in renewable-based islanded AC microgrids. Also, it is observed that the fault current in an islanded microgrid is significantly lesser than the conventional fault current.

To address these limitations, the centralized voltage signal-based fault assessment method is developed by considering the instant changes in magnitude, phase angle and frequency of voltage signal and the RoCoF as features to detect and classify the faults. The HT algorithm is used to extract these features in each fundamental voltage cycle. This fault assessment is a cascaded three-step classifier model where the first classifier is to detect any kind of system disturbances in an islanded AC microgrid—the key advantage of this method is to minimize the nuisance tripping. The second classifier is to differentiate the short circuit faults from nonfault disturbances. Once the detected disturbance is identified as a short circuit fault, the third classifier classifies the fault type for further protection actions. The training performance of all the three classifiers is 95%, 99.8%, and 99%, respectively. Further, the testing performance of the proposed method is validated for the noisy signals in order to guarantee the real system implementation. The fault detection performances are over 100% accuracy for the faults up to 100Ω.

References

[1] A. Hirsch, Y. Parag, J. Guerrero, Microgrids: a review of technologies, key drivers, and outstanding issues, Renew. Sustain. Energy Rev. 90 (2018) 402–411. Available from: https://doi.org/10.1016/j.rser.2018.03.040.

[2] M.A. Khan, A. Haque, V.S.B. Kurukuru, Performance assessment of stand-alone transformerless inverters, Int. Trans. Electr. Energy Syst. (2019). Available from: https://doi.org/10.1002/2050-7038.12156.

[3] M.A. Khan, A. Haque, V.S.B. Kurukuru, M. Saad, Advanced control strategy with voltage sag classification for single-phase grid-connected photovoltaic system, IEEE J. Emerg. Sel. Top. Ind. Electron. (2020). Available from: https://doi.org/10.1109/JESTIE.2020.3041704.

[4] M.A. Khan, A. Haque, K.V.S. Bharath, Hybrid voltage control for stand alone transformerless inverter, in: 2018 2nd IEEE International Conference on Power Electronics, Intelligent Control and Energy Systems (ICPEICES), 2018, pp. 552–557, doi: 10.1109/ICPEICES.2018.8897341.

[5] H. Nie, Y. Chen, Y. Xia, S. Huang, B. Liu, Optimizing the post-disaster control of islanded microgrid: a multi-agent deep reinforcement learning approach, IEEE Access. 8 (2020) 153455–153469. Available from: https://doi.org/10.1109/ACCESS.2020.3018142.

[6] Z. Dehghani Arani, S.A. Taher, A. Ghasemi, M. Shahidehpour, Application of multi-resonator notch frequency control for tracking the frequency in low inertia microgrids under distorted grid conditions, IEEE Trans. Smart Grid 10 (1) (2019) 337–349. Available from: https://doi.org/10.1109/TSG.2017.2738668.

[7] M. Brenna, G.C. Lazaroiu, G. Superti-Furga, E. Tironi, Bidirectional front end converter for DG with disturbance insensitivity and islanding-detection capability, IEEE Trans. Power Deliv. 23 (2) (2008) 907–914. Available from: https://doi.org/10.1109/TPWRD.2007.915997.

[8] G. Magdy, G. Shabib, A.A. Elbaset, Y. Mitani, A novel coordination scheme of virtual inertia control and digital protection for microgrid dynamic security considering high renewable energy penetration, IET Renew. Power Gener. 13 (3) (2019) 462–474. Available from: https://doi.org/10.1049/iet-rpg.2018.5513.

[9] L. Sigrist, A UFLS scheme for small isolated power systems using rate-of-change of frequency, IEEE Trans. Power Syst. 30 (4) (2015) 2192–2193. Available from: https://doi.org/10.1109/TPWRS.2014.2357218.

[10] M.A. Azzouz, A. Hooshyar, E.F. El-Saadany, Resilience enhancement of microgrids with inverter-interfaced DGs by enabling faulty phase selection, IEEE Trans. Smart Grid 9 (6) (2018) 6578–6589. Available from: https://doi.org/10.1109/TSG.2017.2716342.

[11] H.J. Laaksonen, Protection principles for future microgrids, IEEE Trans. Power Electron. 25 (12) (2010) 2910–2918. Available from: https://doi.org/10.1109/TPEL.2010.2066990.

[12] Z. Shuai, et al., Microgrid stability: classification and a review, Renew. Sustain. Energy Rev. 58 (2016) 167–179. Available from: https://doi.org/10.1016/j.rser.2015.12.201.

[13] P. Korba, M. Larsson, C. Rehtanz, Detection of oscillations in power systems using Kalman filtering techniques, in: Proceedings of 2003 IEEE Conference on Control Applications, 2003. CCA 2003, pp. 183–188, doi: 10.1109/CCA.2003.1223290.

[14] S. Mirsaeidi, X. Dong, S. Shi, D. Tzelepis, Challenges, advances and future directions in protection of hybrid AC/DC microgrids, IET Renew. Power Gener. 11 (12) (2017) 1495–1502. Available from: https://doi.org/10.1049/iet-rpg.2017.0079.

[15] M.A. Khan, A. Haque, V.S.B. Kurukuru, Droop based Low voltage ride through implementation for grid integrated photovoltaic system, in: 2019 International Conference on Power Electronics, Control and Automation (ICPECA), Nov. 2019, pp. 1–5, doi: 10.1109/ICPECA47973.2019.8975467.

[16] M.A. Khan, A. Haque, K.V.S. Bharath, Control and stability analysis of H5 transformerless inverter topology, in: 2018 International Conference on Computing, Power and Communication Technologies (GUCON), 2018, pp. 310–315, doi: 10.1109/GUCON.2018.8674915.

[17] M.A. Khan, V.S. Bharath Kurukuru, A. Haque, S. Mekhilef, Islanding classification mechanism for grid-connected photovoltaic systems, IEEE J. Emerg. Sel. Top. Power Electron. 9 (2) (2021) 1966–1975. Available from: https://doi.org/10.1109/JESTPE.2020.2986262.

[18] E. Pashajavid, A. Ghosh, Frequency support for remote microgrid systems with intermittent distributed energy resources—a two-level hierarchical strategy, IEEE Syst. J. 12 (3) (2018) 2760–2771. Available from: https://doi.org/10.1109/JSYST.2017.2661743.

[19] A.A. Renjit, A. Mondal, M.S. Illindala, A.S. Khalsa, Analytical methods for characterizing frequency dynamics in islanded microgrids with gensets and energy storage, IEEE Trans. Ind. Appl. 53 (3) (2017) 1815–1823. Available from: https://doi.org/10.1109/TIA.2017.2657481.

[20] F. Zavoda et al., Power quality in the future grid—Results from CIGRE/CIRED JWG C4.24, in: 2016 17th International Conference on Harmonics and Quality of Power (ICHQP), Oct. 2016, pp. 931–936, doi: 10.1109/ICHQP.2016.7783475.

[21] C. Tu, X. He, Z. Shuai, F. Jiang, Big data issues in smart grid – a review, Renew. Sustain. Energy Rev. 79 (2017) 1099–1107. Available from: https://doi.org/10.1016/j.rser.2017.05.134.
[22] H. Jiang, et al., Big data-based approach to detect, locate, and enhance the stability of an unplanned microgrid islanding, J. Energy Eng. 143 (5) (2017) 04017045. Available from: https://doi.org/10.1061/(ASCE)EY.1943-7897.0000473.
[23] E. Casagrande, W.L. Woon, H.H. Zeineldin, D. Svetinovic, A differential sequence component protection scheme for microgrids with inverter-based distributed generators, IEEE Trans. Smart Grid 5 (1) (2014) 29–37. Available from: https://doi.org/10.1109/TSG.2013.2251017.
[24] P.J. Douglass, K. Heussen, S. You, O. Gehrke, J. Ostergaard, System frequency as information carrier in AC power systems, IEEE Trans. Power Deliv. 30 (2) (2015) 773–782. Available from: https://doi.org/10.1109/TPWRD.2014.2335694.
[25] I.P. Nikolakakos, H.H. Zeineldin, M.S. El-Moursi, N.D. Hatziargyriou, Stability evaluation of interconnected multi-inverter microgrids through critical clusters, IEEE Trans. Power Syst. 31 (4) (2016) 3060–3072. Available from: https://doi.org/10.1109/TPWRS.2015.2476826.
[26] R.M. Kamel, New inverter control for balancing standalone micro-grid phase voltages: a review on MG power quality improvement, Renew. Sustain. Energy Rev. 63 (2016) 520–532. Available from: https://doi.org/10.1016/j.rser.2016.05.074.
[27] L. Cai, N.F. Thornhill, S. Kuenzel, B.C. Pal, Real-time detection of power system disturbances based on \$k\$-nearest neighbor analysis, IEEE Access. 5 (2017) 5631–5639. Available from: https://doi.org/10.1109/ACCESS.2017.2679006.
[28] A. Arunan, J. Ravishankar, E. Ambikairajah, Improved disturbance detection and load shedding technique for low voltage islanded microgrids, IET Gener. Transm. Distrib. 13 (11) (2019) 2162–2172. Available from: https://doi.org/10.1049/iet-gtd.2018.5707.
[29] J.J.Q. Yu, Y. Hou, A.Y.S. Lam, V.O.K. Li, Intelligent fault detection scheme for microgrids with wavelet-based deep neural networks, IEEE Trans. Smart Grid 10 (2) (2019) 1694–1703. Available from: https://doi.org/10.1109/TSG.2017.2776310.
[30] O. Simeone, A brief introduction to machine learning for engineers, Found. Trends® Signal. Process. 12 (3–4) (2018) 200–431. Available from: https://doi.org/10.1561/2000000102.
[31] T.S. Ustun, C. Ozansoy, A. Zayegh, Modeling of a centralized microgrid protection system and distributed energy resources according to IEC 61850-7-420, IEEE Trans. Power Syst. 27 (3) (2012) 1560–1567. Available from: https://doi.org/10.1109/TPWRS.2012.2185072.

CHAPTER EIGHT

Modeling of electric vehicle charging station using solar photovoltaic system with fuzzy logic controller

Priyaratnam[1], Anjali Jain[1], Neelam Verma[1], Rabindra Nath Shaw[2] and Ankush Ghosh[3]

[1]Department of Electrical and Electronics Engineering, Amity University, Noida, India
[2]Department of International Relations, Bharath Institute of Higher Education and Research (Deemed to be University), Chennai, India
[3]School of Engineering and Applied Sciences, The Neotia University, Sarisha, India

8.1 Introduction

According to the present senior of countries, great research has been going on to improve the electric vehicles (EVs) in various fields. Every country is trying to control the growing pollution as well as greenhouse effects and also increasing the prices of crude oils products, the transport sector has determinate a road map for the use of EVs [1] and how to aware the people to use the EV instead of petroleum vehicle. As a result, we use alternative sources of energy, which is renewable energy like wind, solar [2]. It is observed that due to increasing of uses of EVs we need more number of charging stations. For reducing the charging time of EVs we need to set up a fast-charging station. For that we can use a photovoltaic (PV) array system that has a unique optimum operating point is known as maximum power point tracking (MPPT) at which maximum energy/power can be obtained from the PV system [3]. As we change in irradiance value that we provide the input on the PV array then the graphs of maximum power point (MPP) will also vary.

We are well aware that a lot of MPPT methods have been suggested to exploit the maximum possible power from the PV array. We can classify these methods into classical control methods or intelligent control methods. Under the classical control method, like open-loop control and

Applications of AI and IOT in Renewable Energy.
DOI: https://doi.org/10.1016/B978-0-323-91699-8.00008-5

Incremental Conductance [4] and fuzzy and neural control comes under intelligent control.

There are many methods of tracking MPP for PV arrays that have been discussed by Trishan and Chapman [5]. It comprises all the techniques that imply in the field. By comparing the PV array conduction of instantaneous and incremental, Hussein et al. [6] have developed a new MPPT system for tracking a Maximum Power Operating Point. Thulasiyammal and Sutha [7] has introduced the new method for the MPPT that named constant voltage tracking by the breakdown characteristic curve of PV array andalso the operating theory of PV array. The fuzzy logic (FL)-based perturb and observe MPPT control in solar panels was developed by Hussein et al. [6]. In the MATLAB/Simulink, we analyzed solar PV system is considered for supply power to the battery. Simulation results showed that fuzzy logic controller (FLC)-based MPPT has given better performance as compare to the MPPT [8].

In this chapter, modeling and simulation of an EV charging station for DC charging are proposed and formulated with two different controllers. In this chapter, the authors described the system model and MATLAB simulation of the fast-charging station in the following sections. In Section 8.2, a component of charging station and control methods for solar PV charging of EVs and considered in Section 8.3. Simulation and results of the system station discuss in Section 8.4. Finally, conclusions are presented in Section 8.5.

8.2 Components of charging station

The main component of the EV charging station is PV array, DC-DC converter, battery boost converter [9,10] (Fig. 8.1).

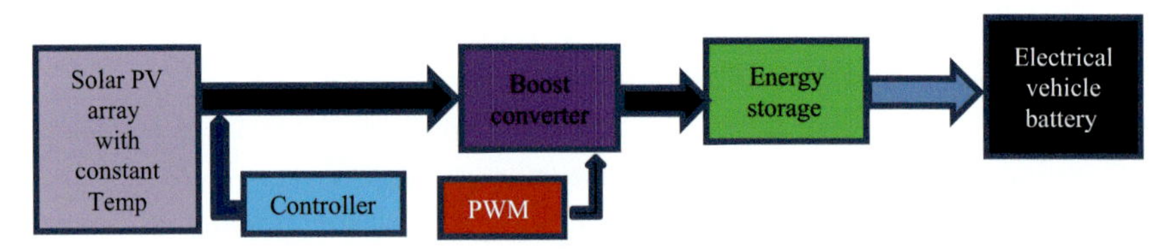

Figure 8.1 Block diagram electrical vehicle charging station.

In the above block diagram of the electrical vehicle charging station (EVCS). The source for EVCS is the solar PV array. With the aid of the boost converter, we step up the power and also add the controller to control the boost converter as the change in the value of irradiance and fixed temperature. Then the output (O/P) of the boost converter is fed into the energy storage system and then connect to the load.

8.2.1 Solar photovoltaic array

The solar photovoltaic array usually uses for the conversion of solar power into electrical power. When the solar radiation excites the Diode junction then the PN junction generates electricity. This value of generation depends on the value of irradiation and temperature that we consider [11,12].

The equation according to the electrical equivalent of Solar panel voltage (SPV) model Fig. 8.2 is given below [13]

$$I_{SPV} = I_{SC} - I_o\left(e^{qv/kT} - 1\right) \tag{8.1}$$

$$V_{OC} = kT/q \ln(I_{SC}/I_o + 1) \tag{8.2}$$

$$P = V_{SPV} \times I_{SPV}$$

$$= V_{SPV}I_{SC} - V_{SPV}I_O(e^{qv/kT} - 1) \tag{8.3}$$

Where, V_{SPV} = Solar panel voltage in volt, I_{SPV} = Solar panel current in ampere, P = Solar power in watt, I_O = Reverse Saturation current (ampere), I_{SC} = SC current of the solar panel (ampere), Q stands for electron charge (C), K = Boltzmann's constant, T = temperature of the panel in Kelvin

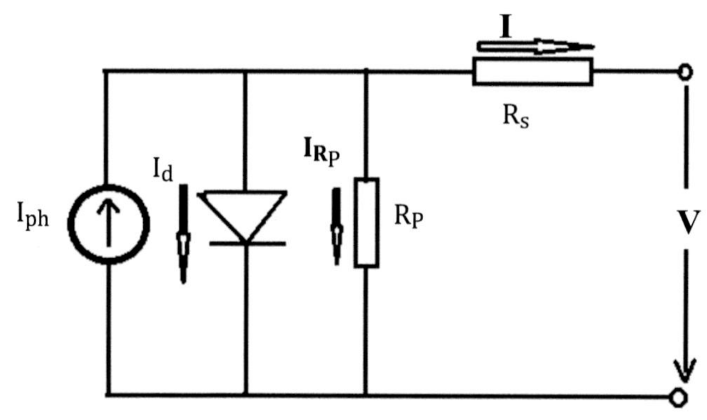

Figure 8.2 The electrical equivalent of solar panel voltage.

Table 8.1 Photovoltaic array specification.

Open circuit voltge V_{oc}	37.14 V
Maximum voltage V_{mp}	30.72
Cells per module (N_{cell})	60
Short circuit current I_{sc}	8 A
Maximum power point current (I_{mp})	7.5 A
Operating frequency	50 Hz

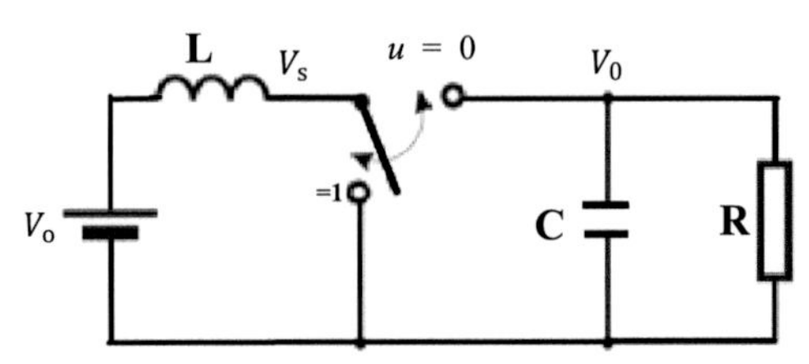

Figure 8.3 Boost converter of a mathematical model.

As we know the value of, Table 8.1

$$q = 1.602 \times 10^{-19} C$$

$$k = 1.381 \times 10^{-23} J/K$$

8.2.2 Boost converter

Generally, a boost converter is called a DC to DC power converter that steps up the voltage from its input to output. So we are using it to step up the voltage that is generated by the solar PV array to the desired leveland it also supplies constant voltage and when we used the MPPT algorithm then it provides maximum power output [14] (Fig. 8.3).

The below-given equation is the mathematical modeling of boost converter:

$$C\frac{dV_C}{dt} = (1-u)i_L - \frac{V_C}{R} - i_O \quad (8.4)$$

$$L\frac{di_L}{dt} = V_{in} - (1-u)V_C \quad (8.5)$$

V_C = Capacitor voltage
i_L = Inductor current
i_O = Resistance
V_{in} = Voltage input
R = Resistance
C = Capacitance

8.2.3 Battery model

In this work, we are using a Lithium-ion battery, but in some work Nickel-Cadmium batteries, is also used. For selecting batteries we are considering some specifications like the nominal voltage, cut-off voltage, energy or nominal energy, ampere-hour capacity, the life of the battery, specific energy, specific power, energy density, charge/discharge cycles of battery, internal resistance, environmental factors, etc [15,16]. The base model of the battery module is Thevenin's equivalent model where open-circuit voltage (V_{oc}) is dependent upon the state of charging (SOC). The battery equivalent circuit is shown in Fig. 8.4.

8.2.4 Battery charger

A battery charger having a bi-directional DC-DC converter with the aid of buck-boost operation. Buck-boost depends upon the switching of the power electronic switch.

Here in Fig. 8.5 if a lower switch is operated then boost action occurs at the left side voltage V_{bat}. Similarly, an upper switch is operated then buck action is occurring in left side voltage V_{bat} [18].

8.3 Control systems strategies

The control system continuously monitors the working of the system at various points and adjusts the input value according to the output value. Here we are using two controller.

1. *Battery charger control system*
2. *PV array control*

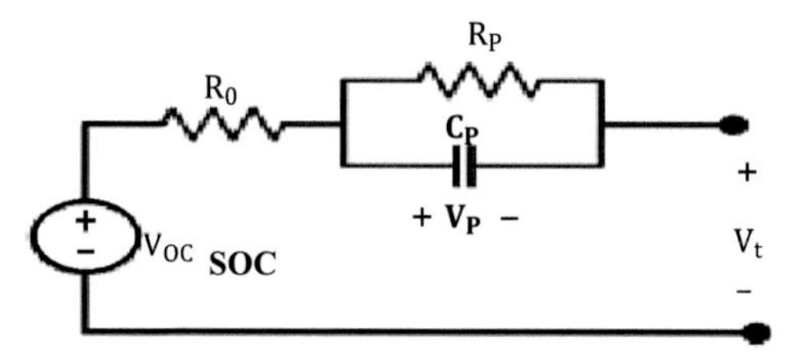

Figure 8.4 Battery equivalent circuit [17].

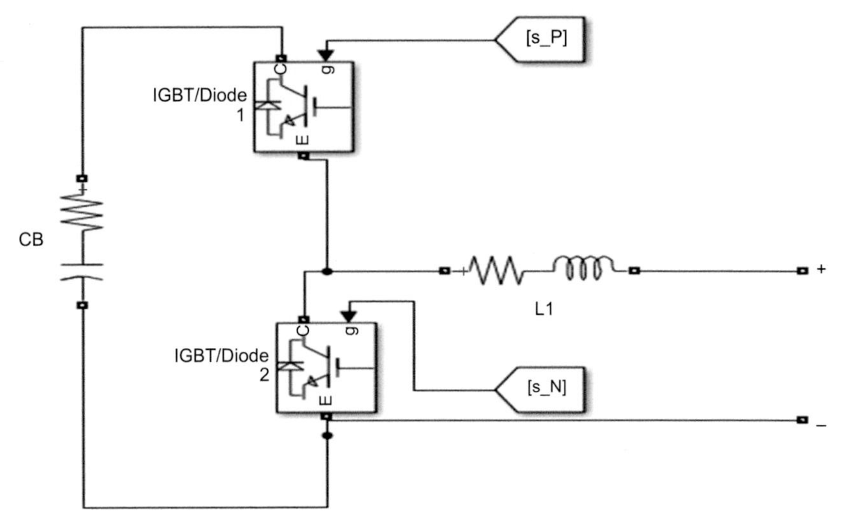

Figure 8.5 Battery charger.

8.3.1 Battery charger control system

In this work, we are using two controllers for battery charging strategies by which EV can be charge "current constant" and "constant voltage." There are two controllers for battery charging strategies by which EV can be charge "constant current" and "constant voltage." Generally at the first stage of charging it is advised to charge the batteries with the aid of constant current strategy and then switch goes to constant voltage strategy. In this work, we are using a constant current strategy in a model of EVCS (Fig. 8.6).

8.3.2 Photovoltaic array control

By many control strategies, we can control the PV array but here we are using two controllers one is MPPT control strategies and the second is fuzzy MPPT control strategies.

As described earlier, the MPP of the system varies with changing the conditions. MPPT is crucial to draw out the maximum power of the PV system. Lately, Intelligent methods are adopted due to their magnificent reasoning, excellent flexibility, and exceptional capability to deal with nonlinear and complex systems [19,20].

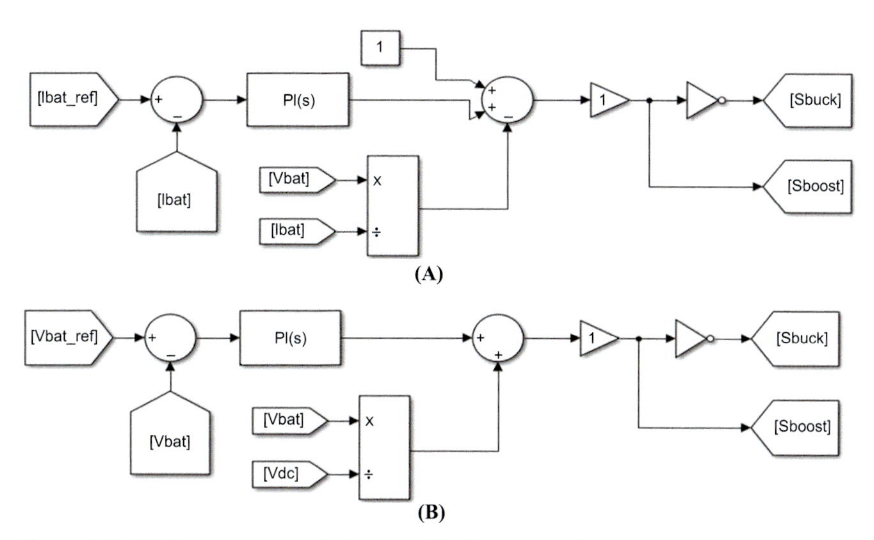

Figure 8.6 (A) Constant current strategy. (B) Constant voltage strategy.

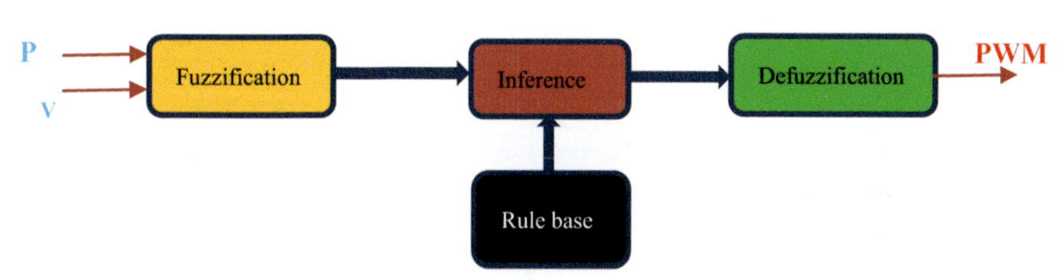

Figure 8.7 Block diagram of fuzzy controller.

3.2.1 Fuzzy logic control

For FL based MPPT controller, we use this instead of the MPPT technique just because it gives a good result as compared to the MPPT technique. The part of an FLC is shown is below give Fig. 8.7 that is fuzzification, inference, rule base, and defuzzification [21].

For checking the MPP tracker for the PV Array system, it introduces to compare the controlling of PV array by the MPPT and FL for different irradiance.

First of all, we have to make fuzzy rules of five membership functionsto track the max power of the PV system under the different irradiance inputs. The input variable we consider for FL is power (*P*) and voltage (*V*) and it gives the PWM [22] (Table 8.2).

Table 8.2 Rule base used in fuzzy logic controller.

Power (P) Voltage (V)	NB	NS	ZE	PS	PB
NB	Positive big (PB)	Positive small (PS)	Zero (ZE)	Negative small (NS)	Negative small (NS)
NS	Positive small (PS)	Positive small (PS)	Zero (ZE)	Negative small (NS)	Negative small (NS)
ZE	Negative small (NS)	Negative small (NS)	Negative small (NS)	Positive big (PB)	Positive big (PB)
PS	Negative small (NS)	Positive big (PB)	Positive small (PS)	Negative Big (NB)	PB
PB	NB	NB	PB	PS	PB

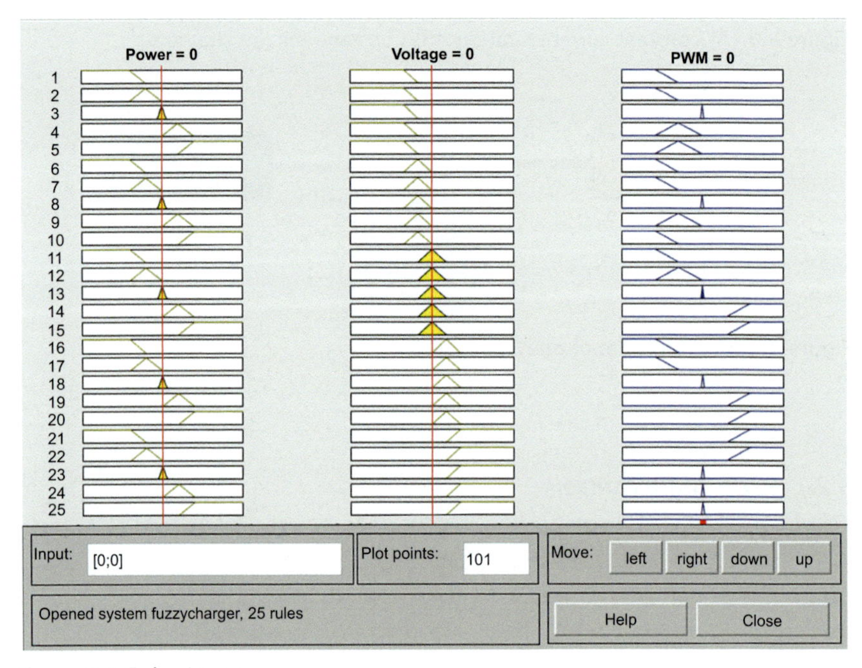

Figure 8.8 Rule viewer.

3.2.2 Rules for fuzzy logic controller

Following are the rules defined for FLC (Figs. 8.8–8.10)

1. If (Power is NB) and (Voltage is NB) then (PWM is NB) (1)
2. If (Power is NS) and (Voltage is NB) then (PWM is NB) (1)
3. If (Power is Z) and (Voltage is NB) then (PWM is Z) (1)

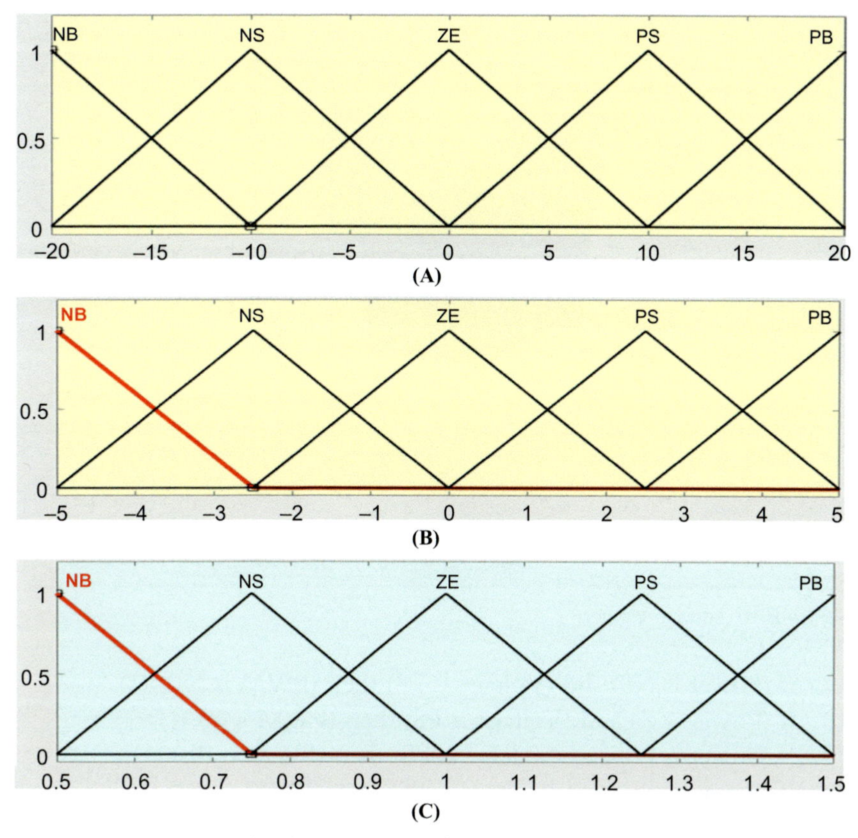

Figure 8.9 (A) Membership function (MF) of the input variable voltage. (B) MF of the input variable power. (C) MFs of the output variable (PWM).

4. If (Power is PS) and (Voltage is NB) then (PWM is NS) (1)
5. If (Power is PB) and (Voltage is NB) then (PWM is NS) (1)
6. If (Power is NB) and (Voltage is NS) then (PWM is NB) (1)
7. If (Power is NS) and (Voltage is NS) then (PWM is NB) (1)
8. If (Power is Z) and (Voltage is NS) then (PWM is Z) (1)
9. If (Power is PS) and (Voltage is NS) then (PWM is NS) (1)
10. If (Power is PB) and (Voltage is NS) then (PWM is NB) (1)
11. If (Power is NB) and (Voltage is Z) then (PWM is NB) (1)
12. If (Power is NS) and (Voltage is Z) then (PWM is NS) (1)
13. If (Power is Z) and (Voltage is Z) then (PWM is Z) (1)
14. If (Power is PS) and (Voltage is Z) then (PWM is PB) (1)
15. If (Power is PB) and (Voltage is Z) then (PWM is PB) (1)
16. If (Power is NB) and (Voltage is PS) then (PWM is NB) (1)

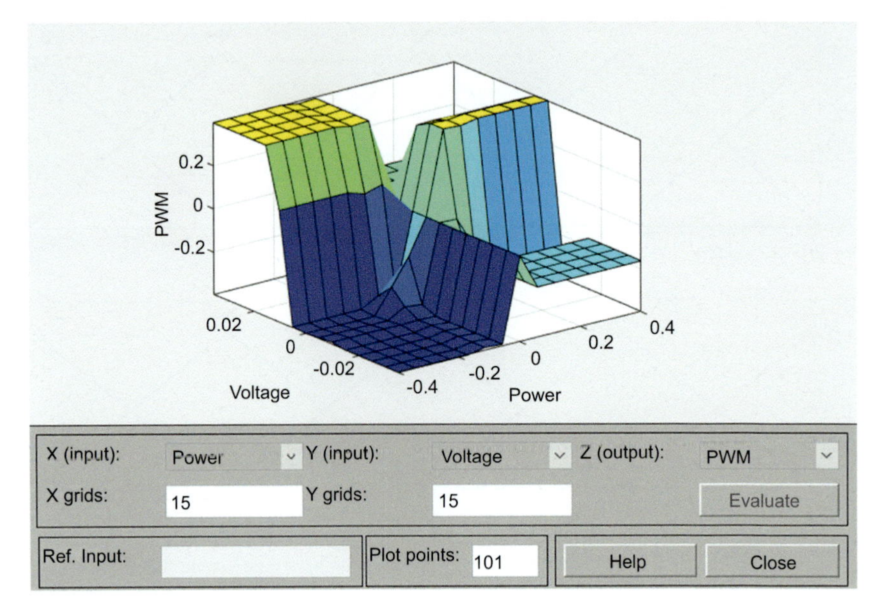

Figure 8.10 Surface viewer.

17. If (Power is NS) and (Voltage is PS) then (PWM is NB) (1)
18. If (Power is Z) and (Voltage is PS) then (PWM is Z) (1)
19. If (Power is PS) and (Voltage is PS) then (PWM is PB) (1)
20. If (Power is PB) and (Voltage is PS) then (PWM is PB) (1)
21. If (Power is NB) and (Voltage is PB) then (PWM is PB) (1)
22. If (Power is NS) and (Voltage is PB) then (PWM is PB) (1)
23. If (Power is Z) and (Voltage is PB) then (PWM is Z) (1)
24. If (Power is PS) and (Voltage is PB) then (PWM is Z) (1)
25. If (Power is PB) and (Voltage is PB) then (PWM is Z) (1)

8.4 Simulation and result

This section gives a brief idea about the system simulation and results on the MATLAB/Simulink diagram. The MATLAB simulation of the EV charging station is represented in Fig. 8.11, where the whole station is connected to a Solar PV array it fed the power to boost the converter with two different controllers MPPT and FLC then it goes to an energy storage system. This energy storage system is used to charge the vehicle battery whenever a vehicle battery is connected to a charging station.

The simulation result is shown in (Figs. 8.12–8.19). where charging of the battery is observed by increasing of SOC where minimum SOC is kept 45% which is increasing as shown in output result of simulation that

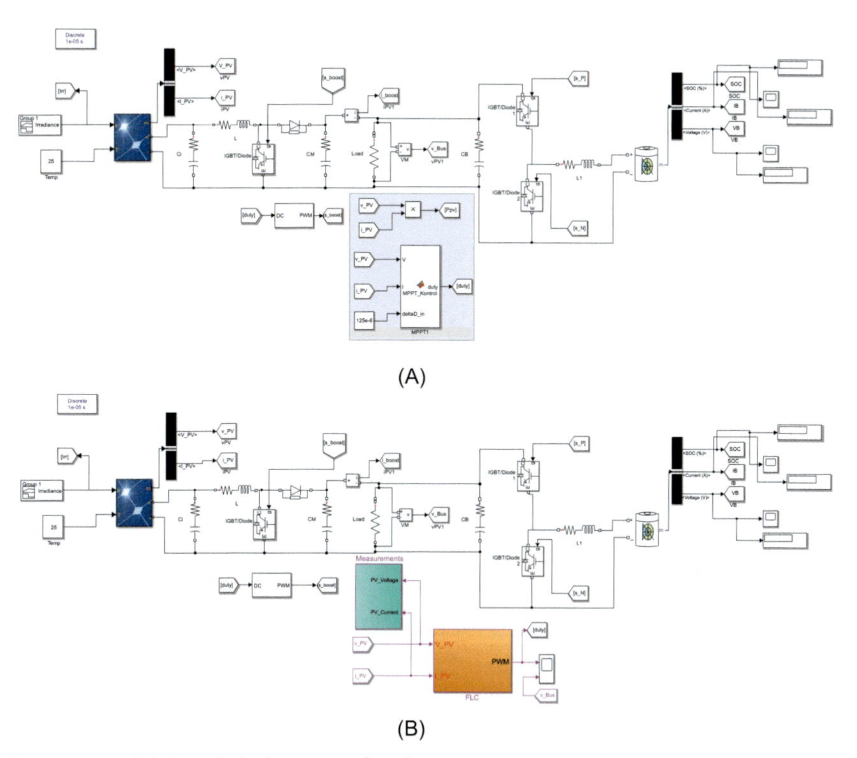

Figure 8.11 (A) Simulink diagram of a charging station with maximum power point tracking controller. (B) Simulink diagram of a charging station with fuzzy logic controller.

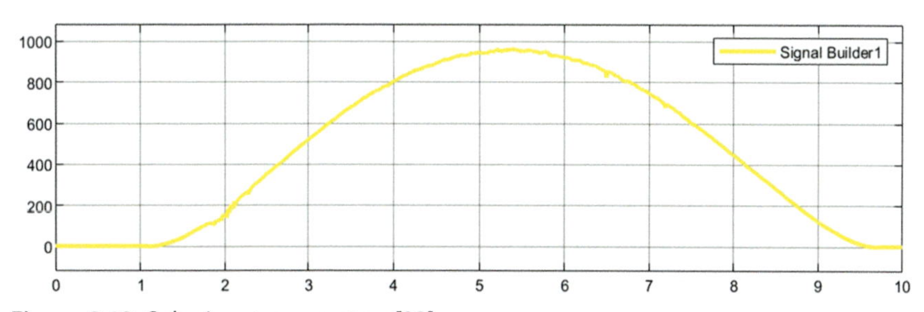

Figure 8.12 Solar input parameters [23].

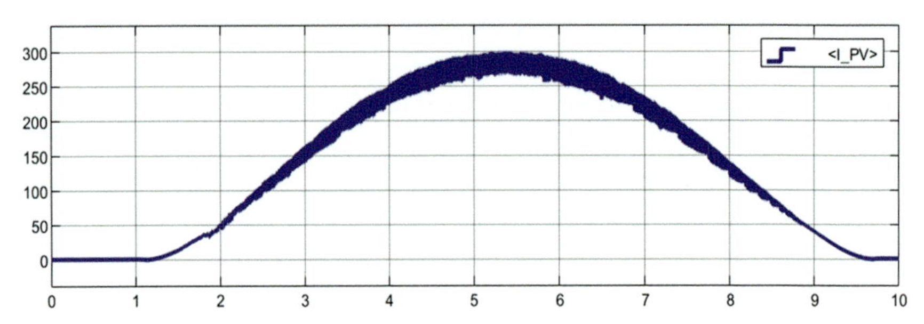

Figure 8.13 Solar photovoltaic output.

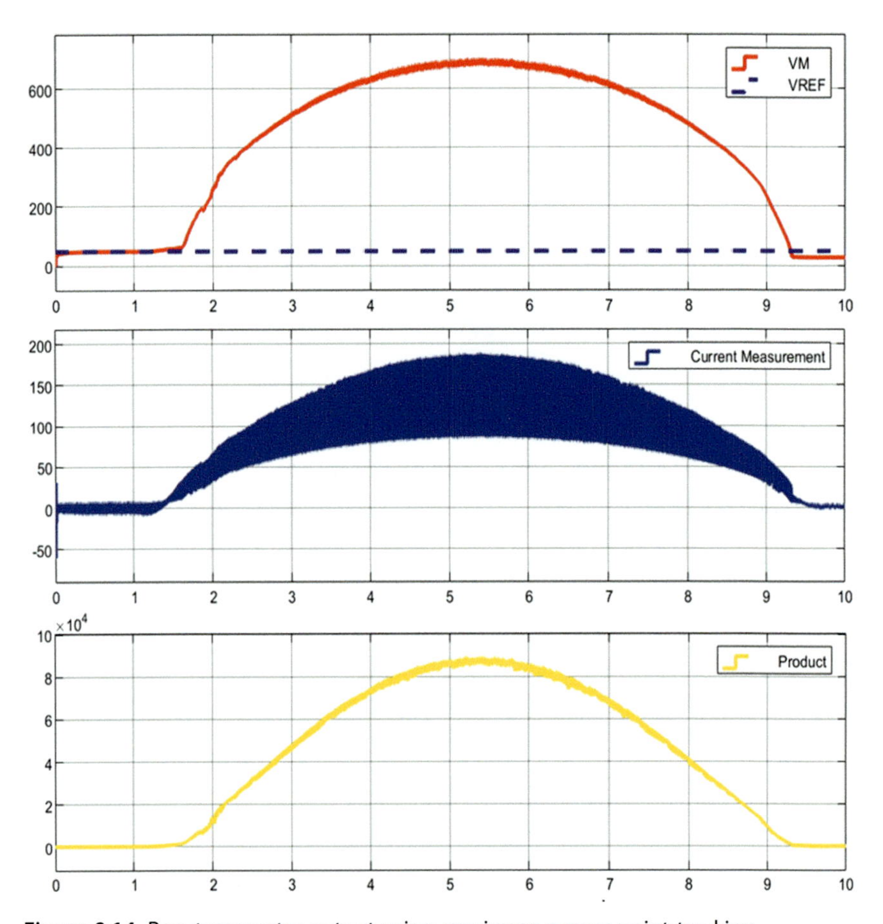

Figure 8.14 Boost converter output using maximum power point tracking.

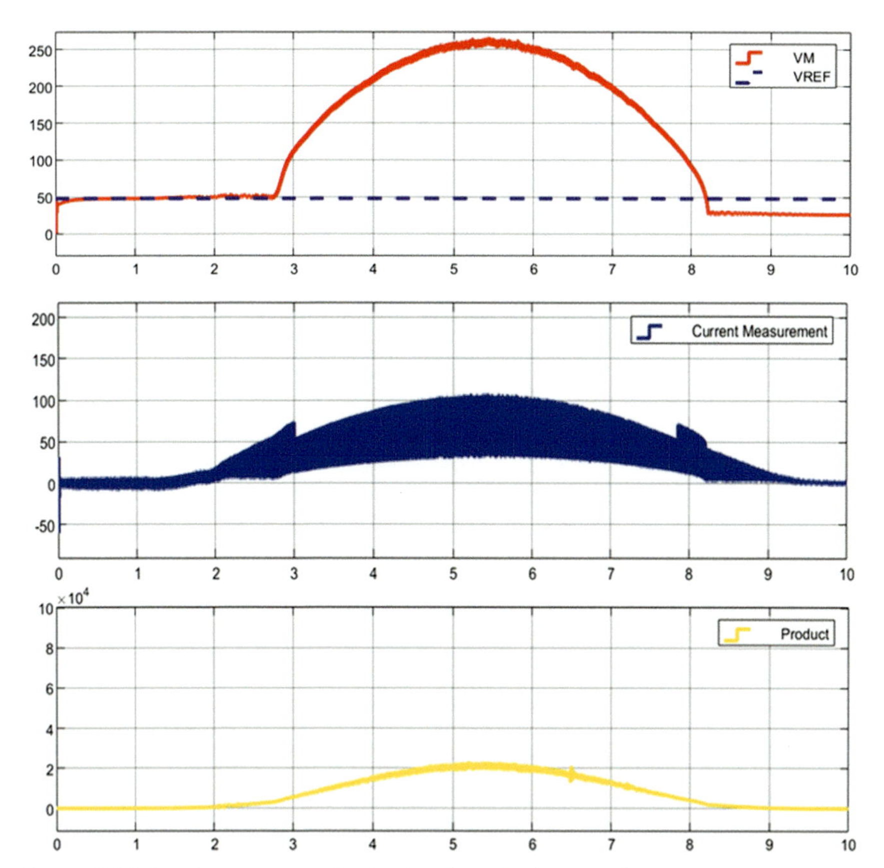

Figure 8.15 Boost converter output using fuzzy logic controller.

shown in Fig. 8.16 and the Fig. 8.18 with different controller MPPT and FLC, respectively (Table 8.3).

Following are the simulation result for the different controllers.

In the above Figs. 8.12 and 8.13, it shows the input and output value of the solar PV array. We got the same graphs for both the controller. Figs. 8.14 and 8.15 show the output using MPPT and FLC, respectively. And Fig. 8.16 shows the SOC of EV using MPPT controller. According to the figure, we can say that first, the battery starts discharging as the time changes, and then it starts charging as the time changes.

According to the graphs of both the controller, we can see in Fig. 8.17 that MPPT gives the maximum power of the system and in Fig. 8.19 FLC also gives the maximum power but powershow in the slope of FLC at every point of the curve. So according to the graph, FLC gives a very appropriate result as compare to MPPT.

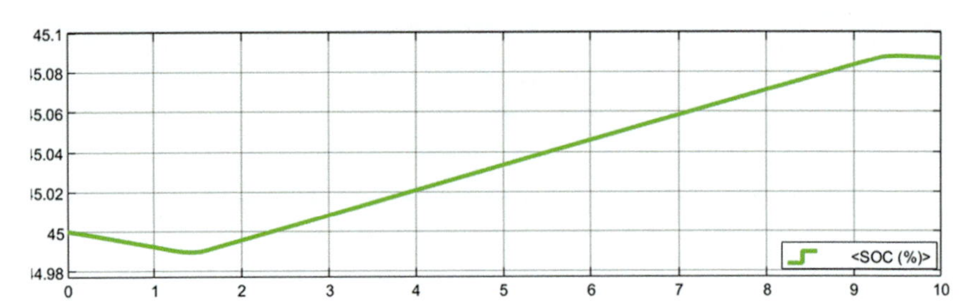

Figure 8.16 Electric vehicle battery state of charging using maximum power point tracking controller.

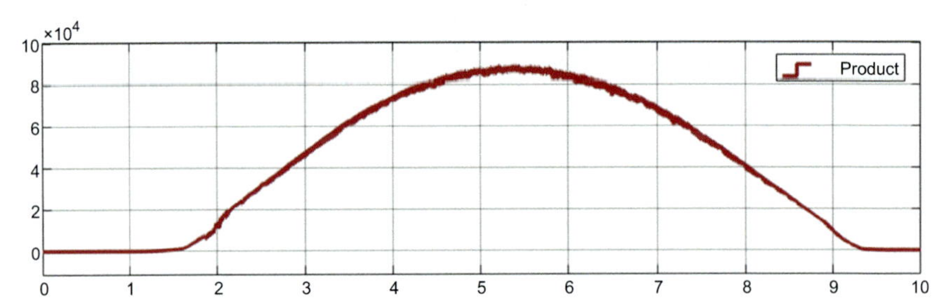

Figure 8.17 Electric vehicle battery power using maximum power point tracking controller.

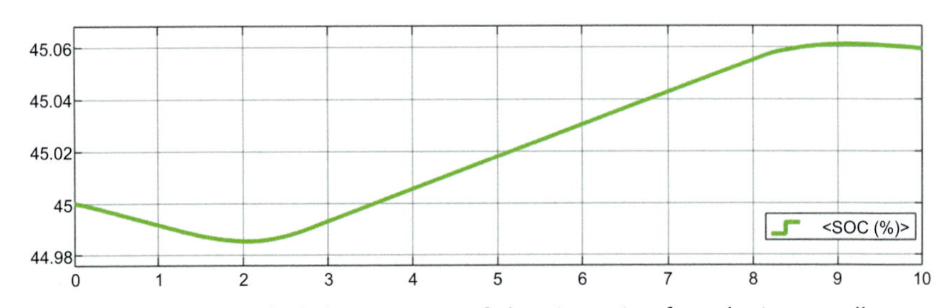

Figure 8.18 Electric vehicle battery state of charging using fuzzy logic controller.

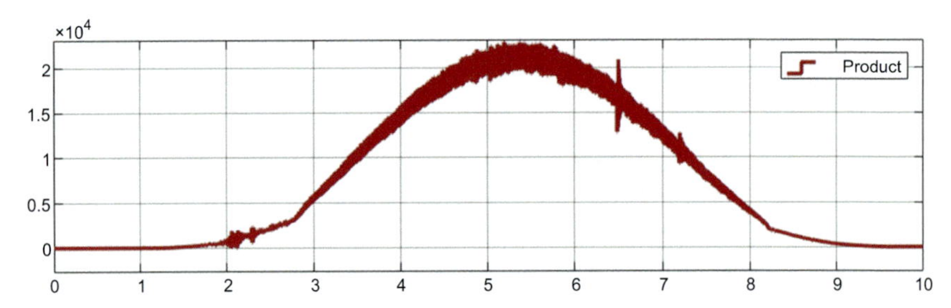

Figure 8.19 Electric vehicle battery power using fuzzy logic controller.

Table 8.3 Battery specification.

Nominal voltage (v)	24
Rated capacity (ah)	50
State of charging	50%
Battery type	Lithium-ion

8.5 Conclusion

This chapter describes a novel approach to charge the EV by renewable energy source. Here solar PV system has been considered and designed the model on MATLAB/Simulink. The controller is used to control the system is MPPT and FLC. Every point of the MPPT can be tracked by using of FLC. Therefore maximum power can be tracked at every point on changing the irradiance which is the most significant parameter for PV cell. Finally a comparison has been done between both the controller technique and the results show that the FL based controller outperform the other.

References

[1] A. Arancibia, K. Strunz, Modeling of an electrical vehicle charging station for fast DC charging, in: Proceedings of the IEEE International Electric Vehicle Conference, 2012, pp. 1–6. <https://doi.org/10.1109/IEVC.2012.6183232>.

[2] P. Singh, S. Bhardwaj, S. Dixit, R.N. Shaw, A. Ghosh, Development of prediction models to determine compressive strength and workability of sustainable concrete with ANN, in: S. Mekhilef, M. Favorskaya, R.K. Pandey, R.N. Shaw (Eds.), Innovations in Electrical and Electronic Engineering. Lecture Notes in Electrical Engineering, vol. 756, Springer, Singapore, 2021. Available from: https://doi.org/10.1007/978-981-16-0749-3_59.

[3] M.A.S. Masoum, H. Dehbonei, E.F. Fuchs, Theoretical and experimental analyses of photovoltaic systems with voltage and current based maximum power point tracking, IEEE Trans. Energy Convers. 17 (4) (2002) 514–522.

[4] A. Jain, A. Mani, A.S. Siddiqui, Cost optimization with electric vehicles and renewable energy sources using priority list method, in: Proceedings of the International Conference on Innovations in Control, Communication and Information Systems (ICICCI), 2017, pp. 1-6, <https://doi.org/10.1109/ICICCIS.2017.8660878>.

[5] E. Trishan, P.L. Chapman, Comparison of photovoltaic array maximum power point tracking techniques, IEEE Trans. Energy Convers. 22 (2) (2007) 439–449. Available from: https://doi.org/10.1109/TEC.2006.874230.

[6] K.H. Hussein, I. Muta, T. Hoshino, M. Osakada, Maximum photovoltaic power tracking: an algorithm for rapidly changing atmospheric conditions, IEE Proc-Gener. Transm. Distrib. 142 (1955) 59–64.

[7] C. Thulasiyammal, S. Sutha, An efficient method of MPPT tracking system of a solar powered uninterruptible power supply application, in: Proceedings of the First

International Conference on Electrical Energy Systems, 2011, pp. 233-236, <https://doi.org/10.1109/ICEES.2011.5725334>.

[8] C.S. Chin, P. Neelakantan, H.P. Yoong, K.T.K. Teo, Fuzzy logic based MPPT for photovoltaic modules influenced by solar irradiation and cell temperature, in: Proceedings of the UkSim Thirteenth International Conference on Computer Modelling and Simulation, 2011, pp. 376-381, <https://doi.org/10.1109/UKSIM.2011.78>.

[9] S.-K. Kim, J.-H. Jeon, C.-H. Cho, J.-B. Ahn, S.-H. Kwon, Dynamic modeling and control of a grid-connected hybrid generation system with versatile power transfer, IEEE Trans. Ind. Electron. 55 (4) (2008) 1677–1688.

[10] R.K. Pachauri, Y.K. Chauhan, Hybrid PV/FC stand alone green power generation: a perspective for indian rural telecommunication systems, in: Proceedings of the International Conference on Issues and Challenges in Intelligent Computing Techniques (ICICT), 2014, pp. 802-810, doi: <https://doi.org/10.1109/ICICICT.2014.6781383>.

[11] S. Silvestre, A. Boronat, A. Chouder, Study of bypass diodes configuration on PV modules, Appl. Energy 86 (9) (2009) 1632–1640.

[12] H.K. Huneria, P. Yadav, R.N. Shaw, D. Saravanan, A. Ghosh, AI and IOT-based model for photovoltaic power generation, in: S. Mekhilef, M. Favorskaya, R.K. Pandey, R.N. Shaw (Eds.), Innovations in Electrical and Electronic Engineering. Lecture Notes in Electrical Engineering, vol. 756, Springer, Singapore, 2021. Available from: https://doi.org/10.1007/978-981-16-0749-3_55.

[13] N. Pragallapati, V. Agarwal, Single phase solar PV module integrated flyback based micro-inverter with novel active power decoupling, in: Proceedings of the Seventh IET International Conference on Power Electronics, Machines and Drives (PEMD 2014), 2014, pp. 1-6, <https://doi.org/10.1049/cp.2014.0269>.

[14] B. Bose, S. Kumar, Ruchira, Design of push-pull flybackconverter interfaced with solar PV system, in: Proceedings of the First International Conference on Power, Control and Computing Technologies (ICPC2T), 2020, pp. 117-121, <https://doi.org/10.1109/ICPC2T48082.2020.9071484>.

[15] Y. Belkhier, A. Achour, R.N. Shaw, A. Ghosh, Performance improvement for PMSG tidal power conversion system with fuzzy gain supervisor passivity-based current control, in: S. Mekhilef, M. Favorskaya, R.K. Pandey, R.N. Shaw (Eds.), Innovations in Electrical and Electronic Engineering. Lecture Notes in Electrical Engineering, vol. 756, Springer, Singapore, 2021. Available from: https://doi.org/10.1007/978-981-16-0749-3_6.

[16] MATLAB Simulink SimPowerSystems 7.6 (R2008a) SimPowerSystems Library Documentation, 2008.

[17] J.Y., Yong, V.K. Ramachandramurty, K.M. Tan, A. Arulampalam, J. Selvaraj, Modelling of electric vehicle fast charging station and impact on network voltage, in: Proceedings of the IEEE Conference on Clean Energy and Technology (CEAT), 2013, pp. 399-404, <https://doi.org/10.1109/CEAT.2013.6775664>.

[18] R.N. Shaw, P. Walde, A. Ghosh, IOT Based MPPT for performance improvement of solar PV arrays operating under partial shade dispersion, in: Proceedings of the IEEE Ninth Power India International Conference (PIICON), 2020, pp. 1-4, <https://doi.org/10.1109/PIICON49524.2020.9112952>.

[19] R.H. Essefi, M. Souissi, H.H. Abdallah, Maximum power point tracking control using neural network for stand-alone photovoltaic syatem, Int. J. Mod. Nonlinear Theory Application 3 (4) (2014) 53–65.

[20] R. Faranda, S. Leva, V. Maugeri, MPPT techniques for PV systems: energetic and cost comparison, in: Proceedings of the IEEE Power and Energy Society General Meeting—Conversion and Delivery of Electrical Energy in the 21st Century, 2008, pp. 1-6, doi: <https://doi.org/10.1109/PES.2008.4596156>.

[21] I.H. Altas, A.M. Sbaraf, A fuzzy logic power tracking controller for a photovoltaic energy conversion scheme, Electr. Power Syst. Res. J. 25 (3) (1992) 227−238.
[22] Gupta, A.; Jain, A., Intelligent control of hybrid power systems for load balancing and levelised cost, in: Proceedings of the IEEE First International Conference on Power Electronics, Intelligent Control and Energy Systems (ICPEICES), 2016, pp. 1-5, doi: <https://doi.org/10.1109/ICPEICES.2016.7853672>.
[23] Signal builder data Files: <https://yadi.sk/d/qCpkYi94VLlKFg>.

CHAPTER NINE

Weather-based solar power generation prediction and anomaly detection

Priyesh Ranjan[1], Pritam Khan[1], Linux Patel[1], Sephali Shradha Khamari[1], Ankush Ghosh[2], Rabindra Nath Shaw[3] and Sudhir Kumar[1]

[1]Department of Electrical Engineering, Indian Institute of Technology, Patna, India
[2]School of Engineering and Applied Sciences, The Neotia University, Sarisha, India
[3]Department of International Relations, Bharath Institute of Higher Education and Research (Deemed to be University), Chennai, India

9.1 Introduction

The traditional generation of energy mainly involves the nonrenewable resources. Nonrenewable resources including coal and petroleum are getting depleted day by day. Burning of fossil fuels pollute our environment thereby causing the jeopardy of global warming. Therefore for eradication of this drawback, the renewable energy sources are the best way out. Renewable sources like wind, hydro, solar, geothermal, and biomass are used extensively for their easily replenishing and eco-friendly nature.

One of the major drawbacks of using renewable energy sources is the erratic nature of the weather. Solar, wind, and hydro are some of the renewable sources of energy that can be easily harnessed. However, prediction of the amount of energy production is a difficult task, given that the weather condition keeps changing. Therefore using the renewable energy resources to meet the required demands commercially is a hurdle based on the estimated production. To overcome this problem, the usage of artificial intelligence (AI) serves as a boon to the present world. The mathematical models that are designed based on the available features predict the power generation before-hand thereby giving an estimate of production to the plant personnel. With the help of AI, energy produced from a plant in a particular interval of time can be easily predicted based on the weather

Applications of AI and IOT in Renewable Energy.
DOI: https://doi.org/10.1016/B978-0-323-91699-8.00009-7

condition. For example, the average power generated by a solar plant can easily be calculated by using a simple regression model based on the available weather and power data. Prediction of the weather can also be carried out by AI and machine learning (ML) based on the available data [1]. Leveraging of AI in turn saves a lot of unrequited manpower.

Physical techniques used in detection are time-consuming and can be inflicted with errors while calculating the response variables [2]. The AI-based models are rigorously tested for the enhancement of their performance and accuracy, thereby increasing their reliability to the users [3,4]. In this work, we predict the solar power generation based on the weather conditions. Regression technique is used to predict the response variables by building a relation between predictor and response variables. One of the advantages of regression is that correlation among the predictor variables is also taken care of precisely and quickly from analysis of variance table. Additionally, anomaly detection in the power generation is also carried out in our work. The actual power generated may differ from the estimated generation owing to the presence of dirt on the solar panels or faulty panel or faulty inverters.

Using regression-based prediction and anomaly detection improves the prediction accuracy over other models like support vector machine (SVM) or auto regressive integrated moving average (ARIMA). Hence, we use the regression model for prediction of solar power generation and anomaly detection in our chapter.

9.1.1 Related work

Multiple works are carried out in the area of predicting the power generation from renewable sources based on features obtained from the climatic conditions. Kothapalli et al. [5] use weather-based features to predict the weather conditions for the next year using ARIMA models. However, the model fails to accurately predict the weather features for time duration of more than two years. A comparative analysis of SVM, Bayesian Enhanced Modified Approach, Ranys method, Broyden—Fletcher—Goldfarb—Shanno, Multi-variant method and Multiple regression for weather prediction is presented in Anusha et al. [6]. The maximum accuracy obtained is limited to 88% which is lower than the proposed work. Krishna [7] uses the spatial and temporal dependencies among the climatic variables with forecasting analysis using an ARIMA model considering the attributes of 97 days. The ARIMA model is only able to provide correct prediction of the weather for the next 15 days as the error values increased proportional to the length of the prediction step. The prediction of electricity demand leveraging a multiple regression analysis on five parameters is carried out in Hsuet al. [8]. Madan et al. [9] use progressive linear regression and SVM for weather forecasting and presents the end result using decision trees. However, the prediction for only the next five days limits the usage of the model. Haida et al. [10] use regression-based daily peak load forecasting method using transformations and reflections. The method yields high absolute error (AE) and is inefficient in summer and winter months. The use of sliding window algorithm for weather forecasting is proposed in Kapoor and Bedi [11]. The method is limited by its poor performance in the months of season change. Furthermore, the authors suggest the use of artificial neural networks to augment the accuracy in future research. Sharifzadeh et al. [12] aim to remove the uncertainties in electrical grids involving the plants based on renewable energy and reaffirms the superiority of artificial neural networks over other ML methods like support vector regression and Gaussian process regression.

9.1.2 Contributions

The major contributions of our work are as follows:

1. Power generation from a solar plant is predicted using regression and high R^2 value is achieved on the available solar power plant data.
2. Anomalous power generation data is found by measuring instances of large deviations of the target label from the mean squared error values obtained through the regression model.
3. Malfunctioning inverters can be identified by measuring the frequency of the occurrence of anomalies for every inverter and comparing with the base line of the available plant data.

The rest of the chapter is organized as follows. In Section 9.2, we discuss the methodology adopted for the regression and anomaly detection on the data from a solar plant and try to identify the malfunctioning inverters based on the occurrences of anomalies. The results are illustrated in Section 9.3. Finally, Section 9.4 concludes the chapter indicating the scope for future work.

9.2 Prediction of solar power generation

The dataset used in this work is collected from a solar power plant in India and is publicly available in Kaggle [13]. The data of power

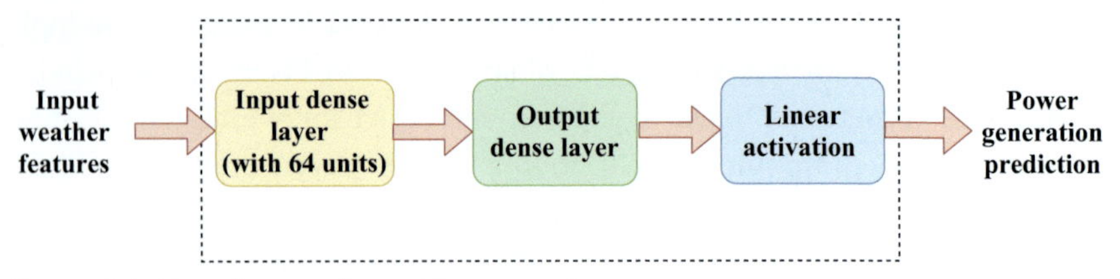

Figure 9.1 Neural network used for regression.

generation in the inverter level and the sensor readings are collected and recorded for a duration of 34 days. In this work, we predict the power generation based on weather features of this dataset leveraging the method of regression [14]. For regression, we leverage a neural network-based model shown in Fig. 9.1, where the model learns from the training data to yield regression results of power generation based on the input features [11]. We also identify anomalies in the power generation data from which we can identify the faulty inverters easily.

9.2.1 Regression-based power generation prediction

We employ the use of regression for the purpose of prediction of the power generation. Regression is one of the supervised ML techniques used to derive a relationship between predictor and response variables. However, in this work we use a deep neural network to obtain the regression outputs because of the possibility of the dataset having nonlinear relation between the input and target variables. Hence, the confusion of appropriateness of linear or nonlinear regression is avoided by leveraging a deep learning model. The regression output can be modeled as:

$$Y = \beta_0 + f(\beta_1, X_1) + f(\beta_2, X_2) + \ldots + f(\beta_p, X_p) + \varepsilon \tag{9.1}$$

where Y is the actual response variable, X_j represents jth distinct predictor variable for $1 \leq j \leq p$, β_j represents jth coefficient of the predictor variable, f is a function applied on the features and ε is the irreducible error term [15].

As we are using a deep neural network for performing the regression task, in our case Y is the target label of the model, X_j is the jth input feature, f is the function applied by the neural network hidden layers and β_j are the weights assigned to each input feature by the neural network [16]. The weather-based features are fed into the input layer of the neural network. The weights of the neural network are analogous to the regression coefficients and these weights get updated during training of the model so as to perform on the test data.

In the neural network-based model, we have one input dense layer of 64 hidden units (number of hidden units is a hyperparameter adjusted during training) and an output dense layer with one neuron as can be observed from Fig. 9.1. The output dense layer has a linear activation. Therefore a single predicted power generation output is obtained corresponding to the input weather features at one time-step and that predicted value is compared to the target variable. The loss function in neural network-based regression is "mean squared error" which evaluates the performance by comparing the output of the model and the target variable and computes the mean value of the squared error for the entire batch. The loss is minimized through back-propagation across the model with the update in the trainable weights. The updated weights make the model ready to be tested on the new set of input features to predict the power generation output. The regression model differs from a classification model in the sense that during classification the labels are qualitative or categorical whereas in our case the labels are quantitative.

The model performance for regression is most commonly deciphered using the R^2 statistic [14]. In other words, R^2 gives a measure of the proportion of variance which always lies between 0 and 1. The value of R^2 is given as:

$$R^2 = \frac{(\mathrm{TSS} - \mathrm{RSS})}{\mathrm{TSS}} = 1 - \frac{\mathrm{RSS}}{\mathrm{TSS}} \tag{9.2}$$

where TSS and RSS are the "Total Sum of Squares," given as the sum of difference of the target values from the average value, and "Residual Sum of Squares," given as the sum of difference between the target and the predicted values, respectively. A higher value of R^2 indicates a model that is able to fit the data well and establish the relation between the input and target variables in a better way. A lower value of R^2 indicates the contrary.

9.2.2 Anomaly in prediction of power generation

The regression values obtained using the model provide an estimate of the target labels with few deviations owing to fluctuations during data acquisition and measurement. However, frequent occurrences of instances of large deviations from the dataset indicate the presence of faulty or malfunctioning equipment. In this work, we detect such anomalous readings leveraging the predicted regression values.

The values predicted from the regression step are compared to the power values of the samples in the dataset and the samples with the target values higher or lower than the regression values are considered to be anomalous by the model. We take a confidence interval of three standard deviations from the mean value as the permissible deviation for a target value. As it is known that for a normal distribution a confidence interval of the first standard deviation represents a confidence interval of 68.2%, the second represents a confidence interval of 95.4% and the third represents a confidence interval of 99.7%, the interval serves well for demarking the anomalous values from the regular values.

Further, multiple instances of anomalous values originating from an inverter may indicate equipment failure or error in data measurement methodology for that inverter. We measure the frequency of anomalous behavior for each inverter and present an analysis of the same.

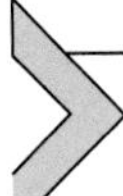

9.3 Experiments and results

9.3.1 Data information

We use the data comprising the AC and DC power generated and the corresponding weather sensor data including features like irradiation, ambient temperature, and module temperature. Power generation data is obtained leveraging the sensors at the inverter level. Each inverter producing the corresponding data is connected to multiple lines of solar panels.

For the purpose of regression, we leverage the weather data and power generation data corresponding to the plant by looking at the common time-steps. The weather features are taken into consideration while the inverter identity features are dropped. This is because of the nonnumerical nature of these features that render them redundant in the regression problem. The daily yield and total yield values are also dropped because of their redundancy in power generation regression as we consider the AC and DC power generated throughout the day. The power generation columns are kept as the target labels that are predicted using the weather features. We normalize each feature vector between 0 and 1 to accelerate the convergence of the model and hence, improve its regression performance.

Fig. 9.2 shows the correlation between every two feature present in the dataset. As observed from Fig. 9.2, every subfigure except the diagonal subfigures represent the scatter plot between the features thereby helping in a bivariate analysis. However, the diagonal subfigures are univariate and hence we obtain the bar-plots indicating the sample distribution for each of them. We observe intuitive correlation between features like irradiation and temperature, which considers an increase in ambient temperature on sunny days, and irradiation and power generation, which highlights the dependency of power generation on the intensity of sunlight incident on the solar plates. We also observe that the features namely total yield and daily yield show poor correlation with other features thereby validating our decision to discard those features.

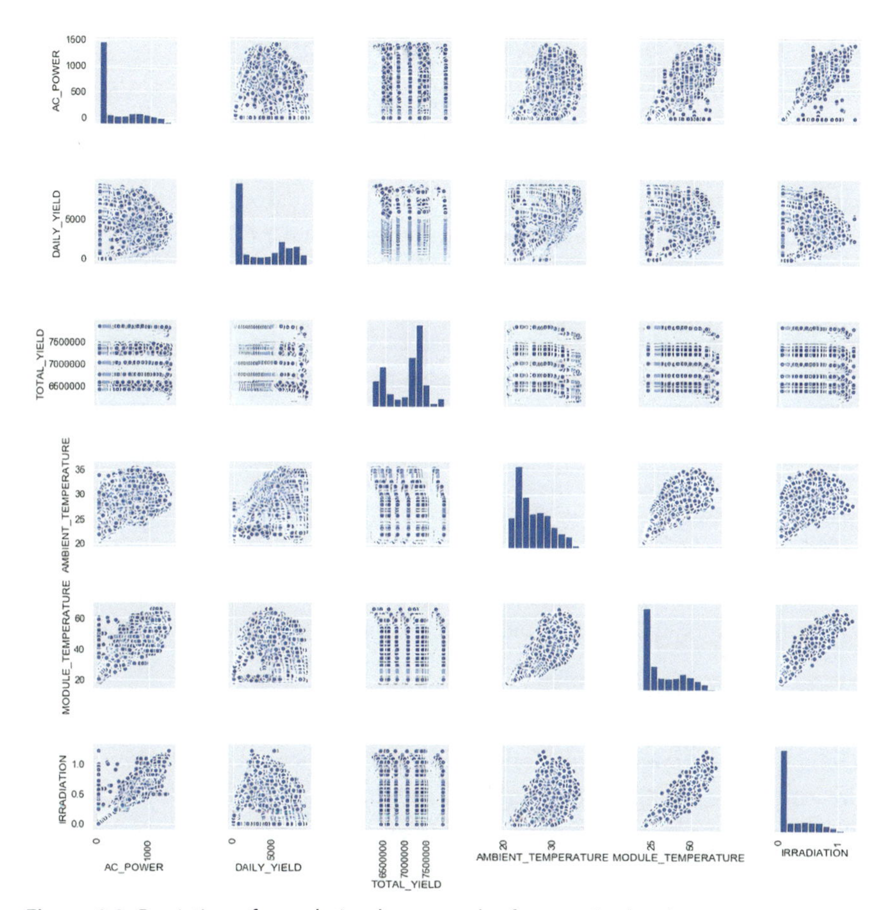

Figure 9.2 Depiction of correlation between the features in the dataset.

We further observe the count of each recurring value for a feature as a bar plot in Fig. 9.2. We observe the high number of 0 values for "IRRADIATION" because of no sunlight at the night time. These 0 values further yield high number of 0 values for AC power as it is directly related to "IRRADIATION." The "DAILY_YIELD" values also showcase a large number of as the values are additive over the entire duration of the day and hence yield 0 till the value of "IRRADIATION" is 0. The lower value of "IRRADIATION" also showcases its effect on the "MODULE_TEMPERATURE" as higher irradiation causes the solar modules to heat up and increase in temperature. Hence, a large number of samples with "IRRADIATION" value 0 results in a large number of 0 values for the "MODULE_TEMPERATURE" feature.

9.3.2 Weather-based power generation prediction

We use a dense network with a single input and output layer with fully connected dense layers. The model contains 64 units in the input dense layer as well as a single neuron in the output dense layer with linear activation. The model is compiled with the mean squared error loss function and the Adam's optimizer. It is trained for a total of 100 epochs with a learning rate of 0.001. We split the data for training and testing in the ratio of 2:1, respectively, thereby validating the performance of our model. We take the values for the last 11 days as the testing samples and the first 23 days are kept for the purpose of training. The R^2 value for the train samples is obtained as 0.9845 and test samples as 0.9666. Further the mean AE value obtained over the entire dataset after regression is 0.02102. This shows the mean value of deviation between the output of the model and the target variable. Fig. 9.3 shows the plot of the predicted and the target labels for the testing data. We observe that the predicted values generated by the model replicate the trend depicted by the target values thereby validating the performance of our model.

9.3.3 Anomaly detection

Using the generated regression values, we predict the anomalous values present in the data by measuring the outliers of the AE. Mathematically, a point having a AE of m is considered as an outlier if:

$$m > \mu \pm 3\sigma \tag{9.3}$$

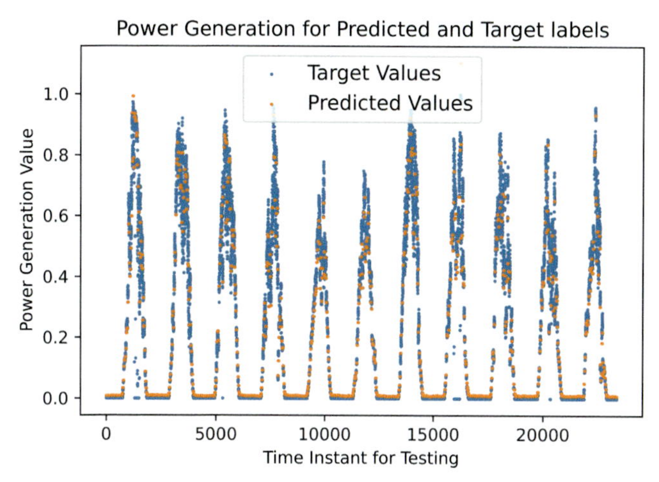

Figure 9.3 Plot depicting the predicted and the target values for the testing samples.

where μ is the average of all AEs of the dataset and σ is the standard deviation of the AEs in the dataset. Only the additive value of the expression $\mu \pm 3\sigma$ is considered, as a smaller AE shows the model lies well in line with the regression output.

Fig. 9.4 depicts the corresponding boxplot [17] for the AE values obtained for the data. At the 99.7th percentile of the AE value, we begin to get the outliers from 0.20078 onwards beginning from the upper end of the box plot in Fig. 9.4. We can also observe the corresponding interval for the nonanomalous data and the value of the median AE.

In this way, we classify the data points into normal and anomalous values and then categorize them based on the inverters producing the corresponding data samples. We obtain a total of 137 anomalous samples in the dataset and categorize them to their corresponding inverters.

The corresponding bar plot illustrating the number of outliers present for each inverter is given in Fig. 9.5. We observe that inverter ID "1BY6WEcLGh8j5v7" gives the highest number of anomalous readings whereas inverters "WRmjgnKYAwPKWDb" and "iCRJl6heRkivqQ3" give no anomalies. However, it can be observed from Fig. 9.5 that most of the inverters have two or more anomalies. Therefore the inverters that yield higher number of anomalies need to be taken care of by the plant personnel as the corresponding inverters tend to hamper the proper functioning of the plant.

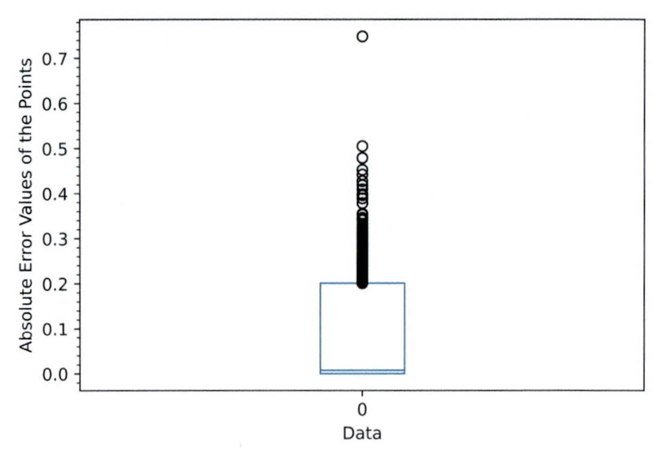

Figure 9.4 Box Plot depicting the 99.7th percentile and the anomalous values.

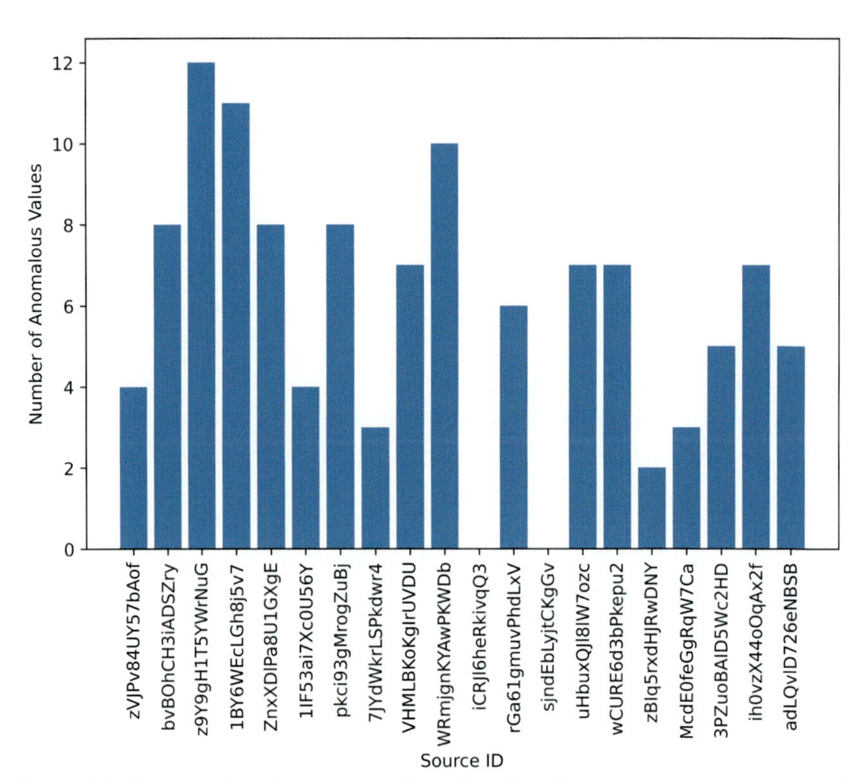

Figure 9.5 Bar plot depicting the number of outliers for every inverter corresponding to the inverter ID.

9.4 Conclusion and future work

In this work we predict the power generation of a solar plant leveraging the weather-based features. The weather features from the solar plant are fed into the deep learning-based regression model to predict the power generation of the solar plants. The R^2 value of the results obtained validate the performance of our deep learning model. We further utilize the results of the regression model to detect the anomalous values in the dataset. The anomalous values are further used to judge the health of the inverters of the solar plant at the generation level.

Future work in the area may include forecasting of power generation by combining the proposed methodology with weather forecasting techniques. Furthermore, it is also possible to use the popular Recurrent Neural Networks or Long Short Term Memory to predict the power generation based on time series datasets. However, similar results are obtained like our regression model at the expense of increased computational complexity.

References

[1] N. Sharma, P. Sharma, D. Irwin, P. Shenoy, Predicting solar generation from weather forecasts using machine learning, in: 2011 IEEE international conference on smart grid communications (SmartGridComm), IEEE, 2011, pp. 528–533.

[2] N. Whitlock, Analyzing solar power energy (IoT analysis), R-bloggers, November 30, 2020. [Online]. Available: https://www.r-bloggers.com/2020/11/analyzing-solar-power-energy-iot-analysis/.

[3] Y. Belkhier, A. Achour, R.N. Shaw, A. Ghosh, Performance improvement for PMSG tidal power conversion system with fuzzy gain supervisor passivity-based current control, in: S. Mekhilef, M. Favorskaya, R.K. Pandey, R.N. Shaw (Eds.), Innovations in Electrical and Electronic Engineering. Lecture Notes in Electrical Engineering, vol. 756, Springer, Singapore, 2021, pp. 81–93. https://doi.org/10.1007/978-981-16-0749-3_6.

[4] P. Singh, S. Bhardwaj, S. Dixit, R.N. Shaw, A. Ghosh, Development of prediction models to determine compressive strength and workability of sustainable concrete with ANN, in: S. Mekhilef, M. Favorskaya, R.K. Pandey, R.N. Shaw (Eds.), Innovations in Electrical and Electronic Engineering. Lecture Notes in Electrical Engineering, vol. 756, Springer, Singapore, 2021, pp. 753–769. https://doi.org/10.1007/978-981-16-0749-3_59.

[5] S. Kothapalli, S.G. Totad, A real-time weather forecasting and analysis, in: 2017 IEEE International Conference on Power, Control, Signals and Instrumentation Engineering (ICPCSI), IEEE, 2017, pp. 1567–1570.

[6] N. Anusha, M.S. Chaithanya, G.J. Reddy, Weather prediction using multi linear regression algorithm, IOP Conf. Ser. Mater. Sci. Eng. 590 (2019) 012034.

[7] G.V. Krishna, An integrated approach for weather forecasting based on data mining and forecasting analysis, Int. J. Computer Appl. 120 (11) (2015) 26–29.

[8] J.-F. Hsu, J.-M. Chang, M.-Y. Cho, Y.-H. Wu, W.-Y. Chang, C.-T. Wang, Development of regression models for prediction of electricity by considering prosperity and climate, in: 2016 3rd International Conference on Green Technology and Sustainable Development (GTSD), 2016, pp. 112-115, doi: 10.1109/GTSD.2016.35.
[9] S. Madan, P. Kumar, S. Rawat, T. Choudhury, Analysis of weather prediction using machine learning & big data, in: 2018 International Conference on Advances in Computing and Communication Engineering (ICACCE), 2018, pp. 259-264, doi: 10.1109/ICACCE.2018.8441679.
[10] T. Haida, S. Muto, Regression based peak load forecasting using a transformation technique, IEEE Trans. Power Syst. 9 (4) (1994) 1788−1794.
[11] P. Kapoor, S.S. Bedi, Weather forecasting using sliding window algorithm, Int. Sch. Res. Notices. 2013 (2013). Available from: https://doi.org/10.1155/2013/156540. Article ID 156540, 5 pages.
[12] M. Sharifzadeh, A. Sikinioti-Lock, N. Shah, Machine-learning methods for integrated renewable power generation: a comparative study of artificial neural networks, support vector regression, and gaussian process regression, Renew. Sustain. Energy Rev. 108 (2019) 513−538.
[13] A. Kannal, Solar power generation data- solar power generation and sensor data for two power plants. [Online]. Available: https://www.kaggle.com/anikannal/solar-power-generation-data.
[14] G. James, D. Witten, T. Hastie, R. Tibshirani, An Introduction to Statistical Learning, Springer, 2013.
[15] L. Zhang, Z. Shi, M.-M. Cheng, Y. Liu, J.-W. Bian, J.T. Zhou, et al., Nonlinear regression via deep negative correlation learning, IEEE Trans. Pattern Anal. Mach. Intell. 43 (3) (2021) 982−998.
[16] Y. Xu, J. Du, L.-R. Dai, C.-H. Lee, A regression approach to speech enhancement based on deep neural networks, IEEE/ACM Trans. Audio, Speech, Language Process. 23 (1) (2015) 7−19. Available from: https://doi.org/10.1109/TASLP.2014.2364452.
[17] J.W. Tukey, Exploratory Data Analysis, Addison-Wesley Publishing Company, Reading, MA, 1977.

CHAPTER TEN

RMSE and MAPE analysis for short-term solar irradiance, solar energy, and load forecasting using a Recurrent Artificial Neural Network

Nilesh Kumar Rai[1], Saravanan D.[1], Labh Kumar[1], Pradum Shukla[1] and Rabindra Nath Shaw[2]

[1]Department of Electrical, Electronics and Communication Engineering, Galgotias University, Greater Noida, India

[2]Department of International Relations, Bharath Institute of Higher Education and Research (Deemed to be University), Chennai, India

10.1 Introduction

One of the biggest challenges of the energy community is to mitigate the band gap between increasing energy demand and rapid growth in population. To overcome this, adopting renewable energy has become a necessary source of energy [1]. Increased conventional sources of energy generation are not a feasible option in view of GHG (greenhouse gas) emission [2]. Photovoltaic (PV) energy generation is nature's gift with the benefits of GHG emission reduction [3], energy demand balancing, minimal maintenance, and the possibility of distributed generation [4]. The substantial growth of PV system due to price drops in the market is significant. Global PV system installations must double every year to meet the energy demand [5]. Solar PV energy generation is predominant to achieving energy demand goals by 2035. The increase in field efficiency of PV systems and noted price reductions have attracted the global community. Recently, PV modules have shown an increase in efficiency from 24% to 30% [6–8]. Though many challenges are associated with PV energy generation, this is likely to be overcome with the motive of GHG reduction. To utilize the luminosity of the sun efficiently, PV systems are automated

Applications of AI and IOT in Renewable Energy.
DOI: https://doi.org/10.1016/B978-0-323-91699-8.00010-3

with tracking mechanisms. The Artificial Neural Network (ANN)-based energy prediction model is an add on feature for forecasting.

10.2 Literature survey

10.2.1 Load forecasting

This technique helps to predict load demand, which tends to balance energy demand and supply. In general, this technique is effective for economic planning and making operational decisions [9]. On the basis of duration, load forecasting is classified as short term, medium term, and long term [10]. Very short term load forecasting study was carried in machine production companies to verify accuracy based on time-based historical data [11]. Electrical load forecasting was carried out between temperature and power consumption based on using ANN and predictions derived with MAPE [12]. Several benchmark studies were compared with AEMO (Australia Energy Market Operator) and resulted in high accuracy in short-term load forecasting [13]. AWNN (Adaptive Wavelet Neural Network)-based forecasting was carried out, which breaks the data into different occurrences and forecasting is distinctly carried out [14]. A combined DWT (Discrete Wavelet Transform) and BNN (Backpropagation Neural Network) were used for load forecasting with significant improvements in accuracy [15].

10.2.2 Solar irradiance forecasting

This technique helps to predict the amount of solar irradiation. The conversion of solar energy to electrical energy depends upon two mechanisms: solar thermal and PV [16]. Irradiance on the earth's surface further depends on the slope of the equator, the height of the sun above the horizon, and atmospheric conditions.

Advanced machine learning techniques were used in wrapped solar irradiance forecasting [17]. A study was carried out for solar irradiance prediction from a few hours to few days using various machine learning models like FoBa, leap Forward, spike slab, Cubist, and bag Earth GCV [18]. Two linear models, ARMA and GARCH, used past data for probabilistic forecasting of solar irradiance [19]. Mixed wavelet neural models were used for high accuracy for prediction of short-term solar irradiance forecasting [20]. In the literature, One day-ahead solar irradiation was predicted using least squares support vector machine technique of specific location with 99.294% accuracy [21]. An online software-based system, SISIFO, was developed to predict solar irradiation by using 3-year global horizon radiation data [22]. A study was carried out for global horizontal irradiance (GHI) based on numerical weather prediction (NWP) [23].

10.2.3 Solar energy forecasting

This mechanism was used for predicting the amount of energy generated by a PV system. Output characteristics of PV systems are highly reliant on irradiation and temperature. Also, other environmental factors like partial shading, soiling effect, location and technical factors like mechanical and electrical tracking techniques, bypass topology and tilt angle are important [24]. A 30–300 min ahead solar irradiation prediction was carried out with an ANN-based system, and RMSE shows significant variation between measured and testing values [25]. An annual energy production forecasting model was developed and studied for virtual mode of access for nonexperts [26]. A reliable Kalman filter-based solar irradiation prediction study was carried out to predict 1 min to 1 h ahead [27].

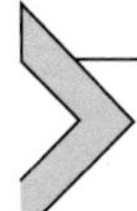

10.3 Prediction methodology

Fig. 10.1

10.4 Artificial Neural Network

It is a computational model similar to the human brain, and it is capable of recognizing any pattern and solving challenges with different learning techniques. The highly interconnected neurons of the ANN system are called nodes. It has three layers, an input layer, hidden layer, and output layer, which are interconnected with each other [28–30]. The input is applied to input layer neurons which transfer the input data to the hidden layer through synapses which contain weight. The weight in synapses tends to change based on learning technique to attain the desired output, where the processed information is transferred to the output layer [31] (Figs. 10.2–10.4).

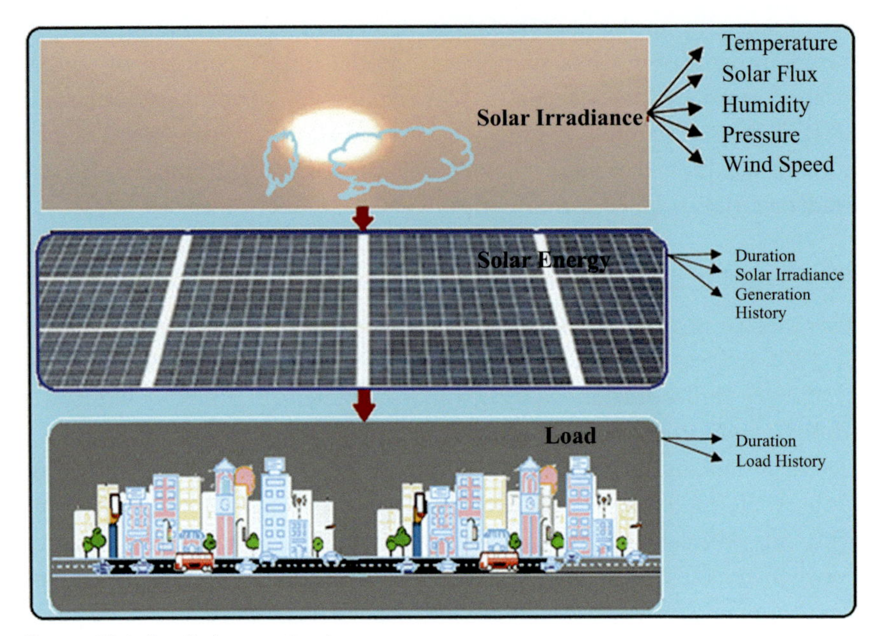

Figure 10.1 Prediction methods.

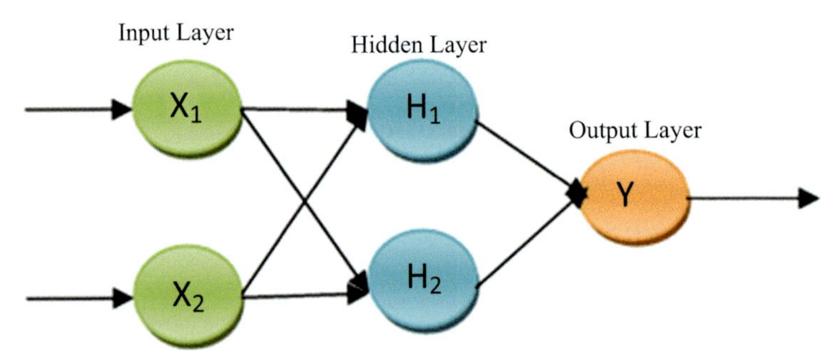

Figure 10.2 Simple ANN model.

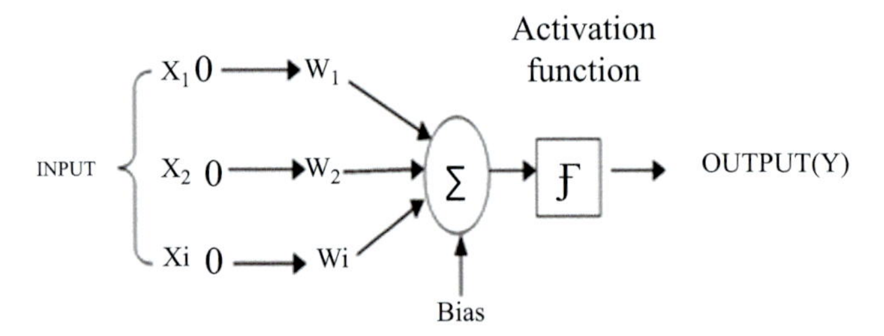

Figure 10.3 Nonlinear model of neurons.

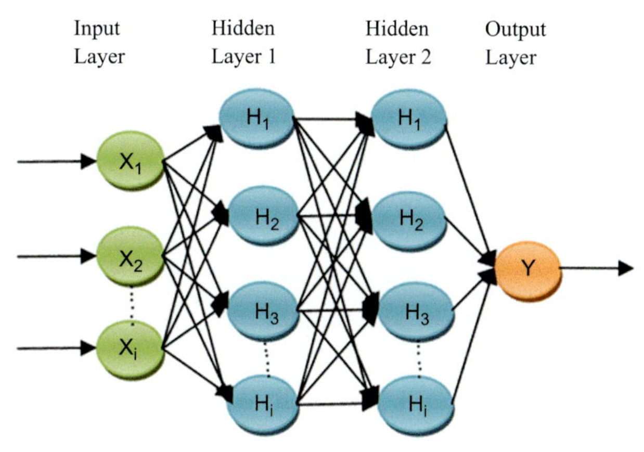

Figure 10.4 Multilayer ANN model.

Table 10.1 Forecasting models, input and output data for RANN algorithm.

Forecasting models	Input data	Output data
Solar irradiance	• Temperature • Wind speed • Humidity • Pressure • Wind direction degrees • Solar flux	• Irradiation
Solar energy	• Duration • Solar irradiance • Generation history	• Generation
Load	• Duration • Load history	• Load

10.5 Data description

In this chapter, various input/output data related to load forecasting, solar irradiance, and solar energy prediction are used (Table 10.1).

10.6 Key performance indicator

RMSE (root mean square error) is a standard mathematical metric used to measure model performance in weather, air quality, and various climatic research [32,33].

The RMSE is defined as:

$$\text{RMSE} = \sqrt{\frac{\sum_{i=1}^{N}(x_i - y_i)^2}{N}} \quad \forall i$$

where x_i,original data; y_i, predicted data; N, number of samples.

10.7 Results and discussion

An assessment was carried out to predict the solar irradiance, solar energy, and load for future uses. A separate recurrent ANN model was used in each case. The input samples were classified into training, validation, and testing as the model required. The different ratios of classified samples were executed to predict the components with minimum error. The developed RANN model has seven hidden neuron layers between the input and output. The input information is passed linearly to hidden neurons through axons and dendrites, a mathematical computation process along with linear weights is appropriated in the hidden layer, and the response information is transferred to the output layer as a prediction.

Solar Irradiance Prediction: The different ratios of samples were simulated to find the gap between actual and predicted solar irradiation. The RMSE and MAPE errors of trained and validated data sets of three different ratios are shown in Fig. 10.5. The error characteristics of actual and predicted solar irradiance are shown in Fig. 10.6. The deviations are projected into it.

Solar Energy Prediction: Samples like historical irradiance have been split into different ratios to find the gap between the actual and predicted solar

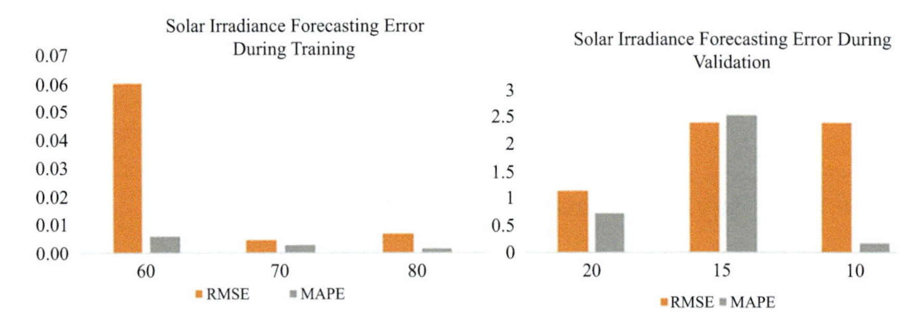

Figure 10.5 Validation and training sample error of solar irradiance forecasting.

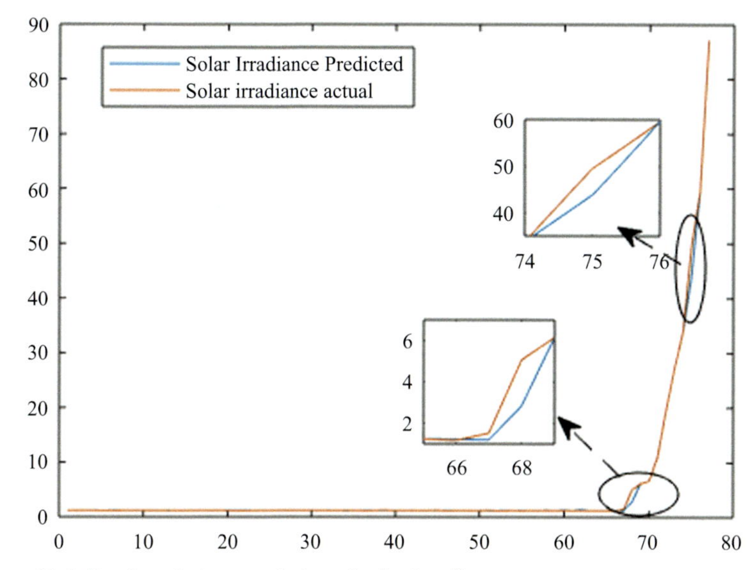

Figure 10.6 Predicted characteristics of solar irradiance.

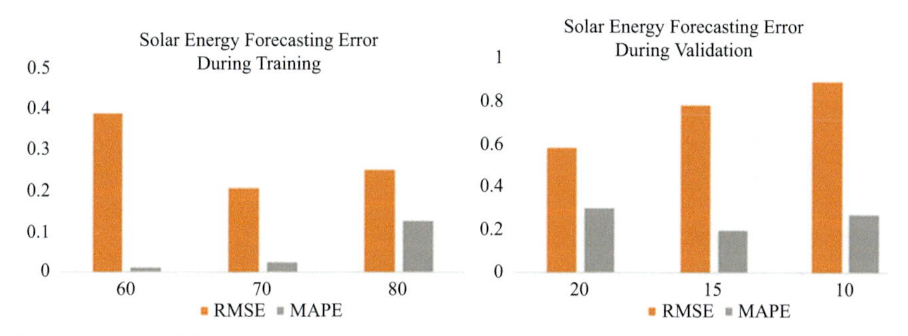

Figure 10.7 Validation and training sample error of solar energy forecasting.

energy. The RMSE and MAPE errors of trained and validated data sets of three different ratios are shown in Fig. 10.7. The deviation characteristics of solar energy forecasting are shown in Fig. 10.8.

Load Prediction: An hourly load history of the month was sampled into three different ratios and applied to develop the RANN system, which reported the RMSE and MAPE errors. Fig. 10.9 shows the RMSE and MAPE errors of trained and validated samples. The accuracy of load prediction is shown in Fig. 10.10 along with deviation.

The developed RANN model predicting the RMSE error of solar irradiance, solar energy, and load forecasting is shown in Table 10.2. The categorized sample data shows a minimum RMSE for 70% of the

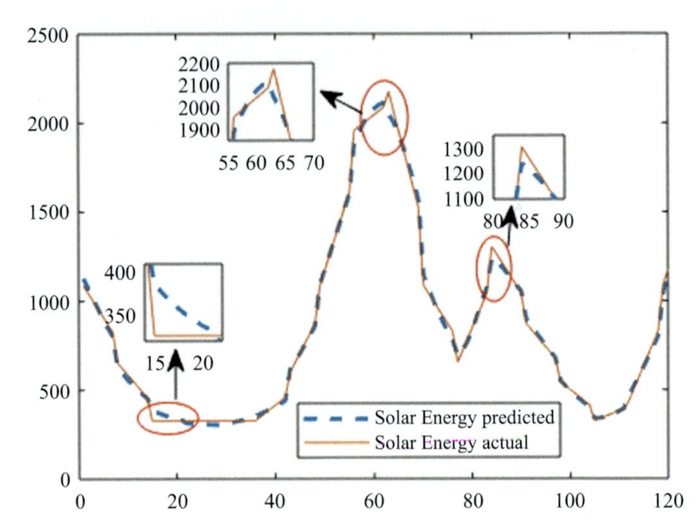

Figure 10.8 Predicted characteristics of solar energy.

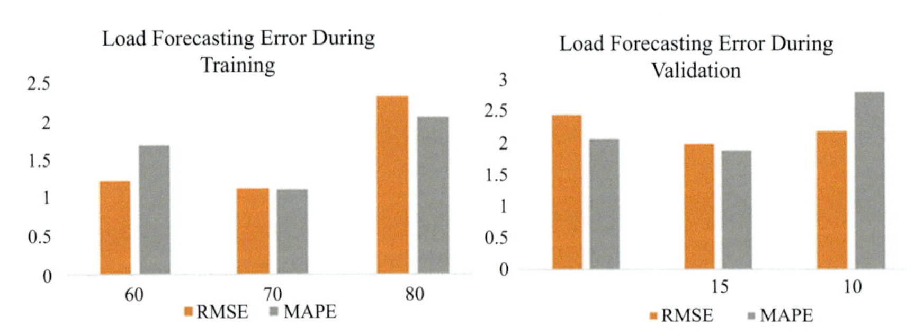

Figure 10.9 Validation and training sample error of load forecasting.

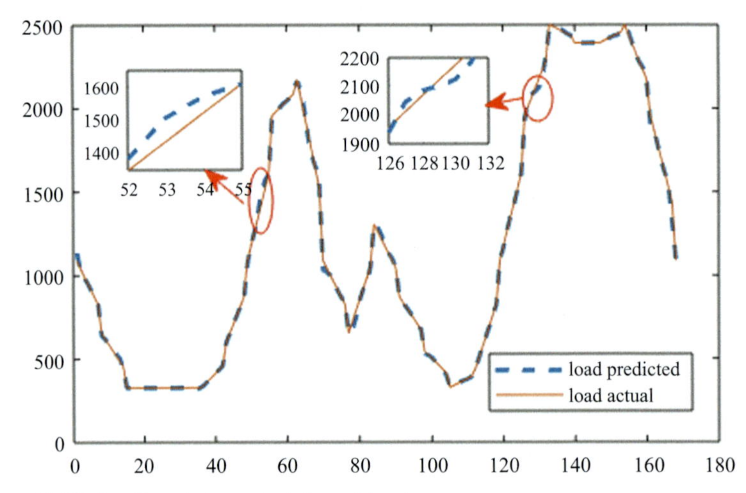

Figure 10.10 Predicted characteristics of load.

Table 10.2 RMSE error predicted from the RANN model.

Forecasting models	**RMSE error in the prediction**		
	60% of samples	**70% of samples**	**80% of samples**
Load	2.3110	0.5250	1.5615
Solar irradiance	1.3633	0.5093	0.2569
Solar energy	2.1545	0.8499	1.793

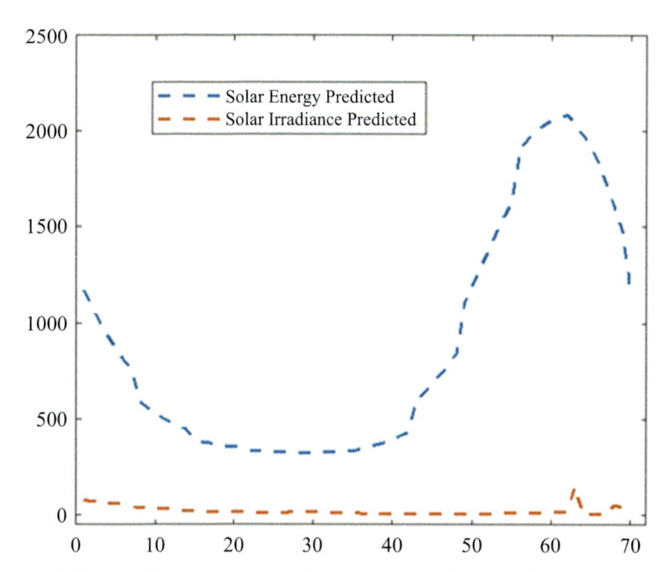

Figure 10.11 Solar irradiance versus solar energy prediction characteristics.

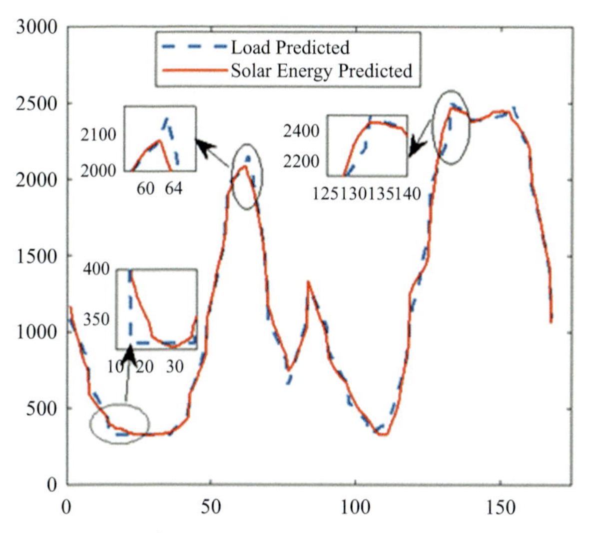

Figure 10.12 Load versus solar energy prediction characteristics.

predicted samples with a ratio of 15% to training and validated samples, respectively.

The solar irradiance and corresponding solar energy generation is projected in Fig. 10.11, where the prediction shows variations from the actual values.

The predicted characteristics of solar energy vs. load is shown in Fig. 10.12. This shows high accuracy for the major prediction duration.

10.8 Conclusions

This study was carried out to bridge economic, energy demand, and GHG emission reductions in terms of forecasting solar irradiance, solar energy, and load for the future. Increasing energy demand and pollution prompts modeling an economic and green energy system with forecasting information. The developed recurrent ANN model results in high levels of accuracy with respect to the predicted values. Future studies can be carried out to reduce the numerical errors in predicting solar irradiance, solar energy, and load by including uncertain and risk inputs into the model.

References

[1] P. Pawar, M. TarunKumar, P. Vittal K, An IoT based Intelligent Smart Energy Management System with accurate forecasting and load strategy for renewable generation, Meas. J. Int. Measurement Confederation 152 (2020). Available from: https://doi.org/10.1016/j.measurement.2019.107187.

[2] What is GHG? | Definition of Greenhouse Gas (GHG) | ADEC ESG Solutions. https://www.esg.adec-innovations.com/about-us/faqs/what-is-ghg/ (accessed 30.01.21).

[3] H.K. Huneria, P. Yadav, R.N. Shaw, D. Saravanan, A. Ghosh, AI and IOT-based model for photovoltaic power generation, in: S. Mekhilef, M. Favorskaya, R.K. Pandey, R.N. Shaw (Eds.), Innovations in Electrical and Electronic Engineering, vol. 756, Springer, Singapore, 2021Lecture Notes in Electrical Engineering. Available from: https://doi.org/10.1007/978-981-16-0749-3_55.

[4] P. Singh, S. Bhardwaj, S. Dixit, R.N. Shaw, A. Ghosh, Development of prediction models to determine compressive strength and workability of sustainable concrete with ANN, in: S. Mekhilef, M. Favorskaya, R.K. Pandey, R.N. Shaw (Eds.), Innovations in Electrical and Electronic Engineering, vol. 756, Springer, Singapore, 2021Lecture Notes in Electrical Engineering. Available from: https://doi.org/10.1007/978-981-16-0749-3_59.

[5] G. Kapoor, V.K. Mishra, R.N. Shaw, A. Ghosh, Fault detection in power transmission system using reverse biorthogonal wavelet, in: S. Mekhilef, M. Favorskaya, R.K. Pandey, R.N. Shaw (Eds.), Innovations in Electrical and Electronic Engineering, vol. 756, Springer, Singapore, 2021Lecture Notes in Electrical Engineering. Available from: https://doi.org/10.1007/978-981-16-0749-3_28.

[6] Y. Belkhier, A. Achour, R.N. Shaw, A. Ghosh, Performance improvement for PMSG tidal power conversion system with fuzzy gain supervisor passivity-based current control, in: S. Mekhilef, M. Favorskaya, R.K. Pandey, R.N. Shaw (Eds.), Innovations in Electrical and Electronic Engineering, vol. 756, Springer, Singapore, 2021Lecture Notes in Electrical Engineering. Available from: https://doi.org/10.1007/978-981-16-0749-3_6.
[7] G. Kapoor, P. Walde, R.N. Shaw, A. Ghosh, HWT-DCDI-based approach for fault identification in six-phase power transmission network, in: S. Mekhilef, M. Favorskaya, R.K. Pandey, R.N. Shaw (Eds.), Innovations in Electrical and Electronic Engineering, vol. 756, Springer, Singapore, 2021Lecture Notes in Electrical Engineering. Available from: https://doi.org/10.1007/978-981-16-0749-3_29.
[8] R.N. Shaw, P. Walde, A. Ghosh, IOT based MPPT for performance improvement of solar PV arrays operating under partial shade dispersion, in: 2020 IEEE 9th Power India International Conference (PIICON) held at Deenbandhu Chhotu Ram University of Science and Technology, SONEPAT, India on Feb 28 – Mar 1 2020.
[9] K. Majumder, K. Chakrabarti, R.N. Shaw, A. Ghosh, Genetic algorithm-based two-tiered load balancing scheme for cloud data centers, in: J.C. Bansal, L.C.C. Fung, M. Simic, A. Ghosh (Eds.), Advances in Applications of Data-Driven Computing, vol. 1319, Springer, Singapore, 2021Advances in Intelligent Systems and Computing. Available from: https://doi.org/10.1007/978-981-33-6919-1_1.
[10] T. Yalcinoz, An educational software package for power systems analysis and operation, Int. J. Elect. Eng. Educ. 42 (4) (2005). Available from: https://doi.org/10.7227/IJEEE.42.4.7.
[11] B. Dietrich, J. Walther, M. Weigold, and E. Abele, "Machine learning based very short-term load forecasting of machine tools," Appl. Energy, 276, 115440, 2020, doi: 10.1016/j.apenergy.2020.115440.
[12] V. Dordonnat, A. Pichavant, A. Pierrot, GEFCom2014 probabilistic electric load forecasting using time series and semi-parametric regression models, Int. J. Forecast. 32 (3) (2016) 1005–1011. Available from: https://doi.org/10.1016/j.ijforecast.2015.11.010.
[13] A.E. Clements, A.S. Hurn, Z. Li, Forecasting day-ahead electricity load using a multiple equation time series approach, Eur. J. Operat. Res. 251 (2) (2016) 522–530. Available from: https://doi.org/10.1016/j.ejor.2015.12.030.
[14] M. Rana, I. Koprinska, Forecasting electricity load with advanced wavelet neural networks, Neurocomputing 182 (2016) 118–132. Available from: https://doi.org/10.1016/j.neucom.2015.12.004.
[15] M. Ghayekhloo, M.B. Menhaj, M. Ghofrani, A hybrid short-term load forecasting with a new data preprocessing framework, Electr. Power Syst. Res. 119 (2015) 138–148. Available from: https://doi.org/10.1016/j.epsr.2014.09.002.
[16] E. Praynlin, J.I. Jensona, Solar radiation forecasting using artificial neural network, 2017. doi:10.1109/IPACT.2017.8244939.
[17] R.N. Shaw, P. Walde, A. Ghosh, Review and analysis of photovoltaic arrays with different configuration system in partial shadowing condition, Int. J. Adv. Sci. Technol. 29 (9s) (2020) 2945–2956.
[18] A. Sharma, A. Kakkar, Forecasting daily global solar irradiance generation using machine learning, Renew. Sustain. Energy Rev. 82 (2018) 2254–2269. Available from: http://doi.org/10.1016/j.rser.2017.08.066.
[19] R.N. Shaw, P. Walde, A. Ghosh, Enhancement of power and performance of 9x4 PV arrays by a novel arrangement with shade dispersion, Test. Eng. Manag. (2020) 13136–13146. ISSN: 0193-4120.
[20] Y. Belkhier, A. Achour, R.N. Shaw, W. Sahraoui, A. Ghosh, Adaptive linear feedback energy-based backstepping and PID control strategy for PMSG driven by a

grid-connected wind turbine, in: S. Mekhilef, M. Favorskaya, R.K. Pandey, R.N. Shaw (Eds.), Innovations in Electrical and Electronic Engineering, vol. 756, Springer, Singapore, 2021Lecture Notes in Electrical Engineering. Available from: https://doi.org/10.1007/978-981-16-0749-3_13.

[21] B.B. Ekici, A least squares support vector machine model for prediction of the next day solar insolation for effective use of PV systems, Meas. J. Int. Measurement Confederation 50 (1) (2014) 255–262. Available from: https://doi.org/10.1016/j.measurement.2014.01.010.

[22] J. Muñoz, O. Perpiñán, A simple model for the prediction of yearly energy yields for grid-connected PV systems starting from monthly meteorological data, Renew. Energy 97 (2016) 680–688. Available from: https://doi.org/10.1016/j.renene.2016.06.023.

[23] J.F. Mejia, M. Giordano, E. Wilcox, Conditional summertime day-ahead solar irradiance forecast, Sol. Energy 163 (2018) 610–622. Available from: https://doi.org/10.1016/j.solener.2018.01.094.

[24] S.M. Awan, Z.A. Khan, M. Aslam, Solar generation forecasting by recurrent neural networks optimized by levenberg-marquardt algorithm, 2018. doi:10.1109/IECON.2018.8591799.

[25] E. Izgi, A. Öztopal, B. Yerli, M.K. Kaymak, A.D. Şahin, Short-mid-term solar power prediction by using artificial neural networks, Sol. Energy 86 (2) (2012) 725–733. Available from: https://doi.org/10.1016/j.solener.2011.11.013.

[26] R.N. Shaw, P. Walde, A. Ghosh, A new model to enhance the power and performances of 4x4 PV arrays with puzzle shade dispersion, Int. J. Innovative Technol. Explor. Eng. 8 (2021) 12.

[27] R.N. Shaw, P. Walde, A. Ghosh, Effects of solar irradiance on load sharing of integrated photovoltaic system with IEEE Standard Bus Network, Int. J. Eng. Adv. Technol. 9 (1) (2019).

[28] S. Agatonovic-Kustrin, R. Beresford, Basic concepts of artificial neural network (ANN) modeling and its application in pharmaceutical research, J. Pharm. Biomed. Anal. (2000). Available from: https://doi.org/10.1016/S0731-7085(99)00272-1.

[29] Artificial Neural Networks-A Study. https://studyres.com/doc/3760182/artificial-neural-networks-a-study (accessed 30.01.21).

[30] M.S. Mhatre, D. Siddiqui, M. Dongre, P. Thakur, A review paper on artificial neural networks: a prediction technique, Int. J. Sci. Eng. Res. (2017).

[31] D. Saravanan, A. Hasan, A. Singh, H. Mansoor, R.N. Shaw, Fault prediction of transformer using machine learning and DGA, 2020. doi:10.1109/GUCON48875.2020.9231086.

[32] Forecast KPIs: RMSE, MAE, MAPE & Bias | by Nicolas Vandeput | Towards Data Science. https://towardsdatascience.com/forecast-kpi-rmse-mae-mape-bias-cdc5703d242d (accessed 30.01.21).

[33] T. Chai, R.R. Draxler, Root mean square error (RMSE) or mean absolute error (MAE)? Arguments against avoiding RMSE in the literature, Geoscientific Model. Dev. (2014). Available from: https://doi.org/10.5194/gmd-7-1247-2014.

CHAPTER ELEVEN

Study and comparative analysis of perturb and observe (P&O) and fuzzy logic based PV-MPPT algorithms

Avinash Kumar Pandey[1], Varsha Singh[2] and Sachin Jain[3]

[1]Electrical Engineering Department, National Institute of Technology, Raipur, Raipur, India
[2]Member IEEE Electrical Engineering Department National Institute of Technology, Raipur, Raipur, India
[3]Senior member IEEE Electrical Engineering Department National Institute of Technology Raipur, Raipur, India

11.1 Introduction

Fuzzy logic control (FLC) is well known artificial intelligence, based control technique [1–3]. It utilizes the prior experience of the functionary about the system to be controlled. The main role of the functionary is to set up decision-based rules by analyzing the system behavior and the linguistic input variables within the framework of the system. The inputs provided to the FLC have to process through three basic stages of fuzzification, decision-making stage, and defuzzification before generating the output [4]. In the fuzzification stage, the input variable is transformed into linguistic variable with the help of predefined membership functions (MFs). The output of the fuzzification stage is then used to generate the fuzzified output according to the rules set defined. Finally, in the defuzzification stage, the fuzzified output is transformed into the required output used for controlling the system. The most interesting fact with respect to the FLC is it does require the exact model of the system during its development. Therefore The FLC has a extensive scope of applications in the field of machine control for systems that have high uncertainty and nonlinearity in their nature [5]. Consequently, the control strategy of FLC is relatively complex and demands high implementation costs.

Nowadays, fuzzy logic-based controllers are getting more popular in the field of photovoltaic (PV) systems [6–8]. The PV source generates

Applications of AI and IOT in Renewable Energy.
DOI: https://doi.org/10.1016/B978-0-323-91699-8.00011-5

electricity from solar energy and shows nonlinear I-V and P-V characteristics [9]. Consequently, the output power is nonuniform for all the operating points and has single maximum power point (MPP) for the given atmospheric condition. Thus the PV source which is directly connected to the load is rare to operate at MPP for all the environmental condition. In addition, the dynamic or fast-changing atmospheric condition continuously changes the operating voltage corresponding to MPP and reduces the performance of the PV system. To tackle variable operating voltage due to changing environmental conditions various PV-maximum power point tracking (PV-MPPT) algorithms are given in the literature for the last 20 years [10–15]. Among the given solutions most classical and frequently used algorithm is perturb and observe (P&O) algorithm [16–18]. The given algorithm is trouble-free and easy to implement. However, it employs the fixed small change in the PV source operating point. It is more suitable for the slow gradual changing environmental conditions. Thus for the fast-changing environmental conditions, it may not be suitable. Here, algorithms using variable step change in the PV operating point may be useful. Such algorithms should be able to predict present operating location with respect to the peak operating point. When the peak operating point is away it may employ a large step change in the PV source operating point to reach near the MPP. And when it reaches near MPP operating it can employ the small step change in the operating point to track exact MPP. Fuzzy logic-based PV-MPPT is one solution for such variable step MPPT algorithm [19,20]. This chapter gives detail of both P&O and fuzzy-based MPPT algorithms.

11.2 Photovoltaic system

The PV power generation system includes four main components: PV module, controlled power interface (DC-DC/DC-AC converters) MPPT controller, and output load as shown in Fig. 11.1. PV Module absorbs photons from the Sunlight and using the Photo-Electric effect converts the solar energy to electrical energy. The generated electrical energy needs to be processed into the required form for their usage. This is typically taken care of by the power electronics converter which is an interface between the PV source and the load. Depending on the load power electronics converters are chosen. Typically, a DC-DC converter is

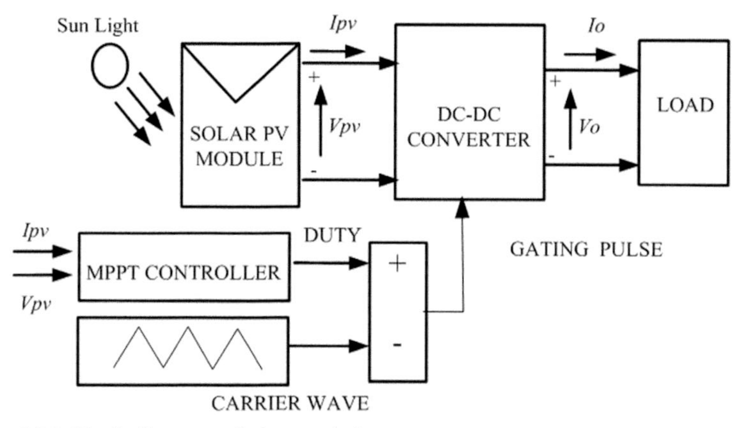

Figure 11.1 Block diagram of photovoltaic system.

employed to operate a PV source at MPP. These converters conditioned the PV source power in such a way that it operates the PV source near MPP. PV source is operated near MPP by MPPT controller which controls the function of the converter. The MPPT controller is required to handle the on/off strategies of PV-fed converters. The MPPT system takes input parameters like PV voltage, PV current ambient temperature, and solar irradiance and provides duty as an output control signal. This control signal is fed to PWM to produce a gating pulse for the converter as can be seen from Fig. 11.1.

11.2.1 Photovoltaic source modeling

PV cells are responsible for generating photocurrent (Iph) from solar energy. The generated photocurrent is the function of incident irradiance and module temperature. Fig. 11.2(A) shows the equivalent circuit of a single diode PV cell. The output cell power is then calculated as voltage times the generated photocurrent. The series-connected PV cells form a PV module [21–24]. This PV module produces nonlinear output power which has a single peak power point. The parameters of the PV module at STC are given in Table 11.1. Fig. 11.2(B) and (C) are I-V and P-V characteristics of PV cells and can be described by the following equations:

$$I_{\text{pv}} = I_{\text{ph}} - I_D - I_p \tag{11.1}$$

$$I_{\text{ph}} = \frac{G}{G_{\text{ref}}}\left(I_{\text{ph,ref}} + \mu_{\text{sc}}.(T - T_n)\right) \tag{11.2}$$

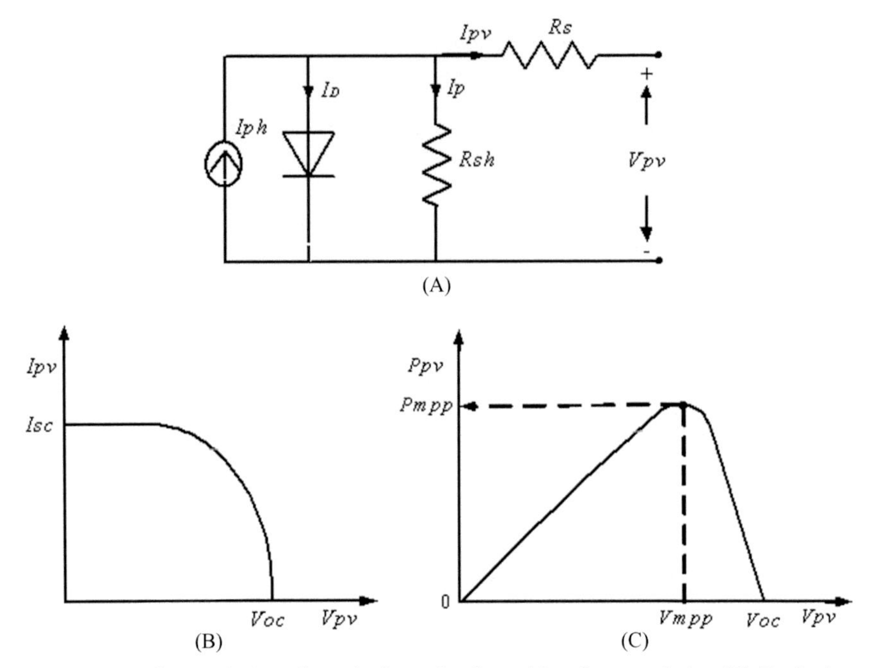

Figure 11.2 Photovoltaic cell equivalent circuit and its characteristics. (A) Equivalent circuit diagram of photovoltaic cell. (B) P-V characteristics. (C) I-V characteristics.

Table 11.1 Parameters of the photovoltaic module at standard test condition.

Parameter	Symbol	Value
Peak power	P_{mpp}	1133 W
Operating voltage at peak power	V_{mpp}	163 V
Current at peak power	I_{mpp}	6.6 A
Open −circuit voltage	V_{oc}	183.5
Short- circuit current	I_{sc}	7.82
Series connected PV cells	N_s	300
Parallel branches	N_p	1

$$I_D = I_0\left(e^{\left(\frac{q}{nk}\frac{(V_{\text{pv}}[-I_{\text{pv}}].R_S)}{T}\right)} - 1\right) \tag{11.3}$$

$$I_p = \frac{V_{\text{pv}} + I_{\text{pv}}.R_s}{R_p} \tag{11.4}$$

Where I_p is current through R_p; I_D is current through the diode; q is electron charge 1.602×10^{-23} C; G is Sunlight irradiance in W/m^2 at STC;

G_{ref} is the referenced irradiance 1000 W/m^2; $I_{ph,ref}$ is the referenced PV current at STC; μ_{sc} is the short circuit current temperature coefficient; T_n is the referenced temperature; n is the ideality factor; T is the temperature in Kelvin; k is the Boltzmann constant; I_0 is reverse saturation current; V_{pv} and I_{pv} are solar panel voltage and current; R_s and R_p are the series and parallel resistances of PV cell, respectively;

11.2.2 DC-DC converter modeling

DC-DC converter plays an important role in PV systems. It is employed as a controlled power interface between the PV module and the load, which controls the power transfer from the source to the load [25,26]. The MPPT system is to address the switching or duty cycle control of the converter, which further provides an appropriate operating voltage of the PV source to operate it at MPP. It is clear from the Eqs. (11.5) and (11.6) that for the constant output voltage of the converter, the source voltage can be adjusted as per the duty ratio. Fig. 11.3 shows the equivalent circuit model of the buck-boost converter. The parameters of model of the buck-boost converter are given in Table 11.2. Where V_O is the load voltage at the output, V_s is the input source voltage, D is duty ratio, t_{on} is on-time of the switch, and t_{off} is off time of the switch.

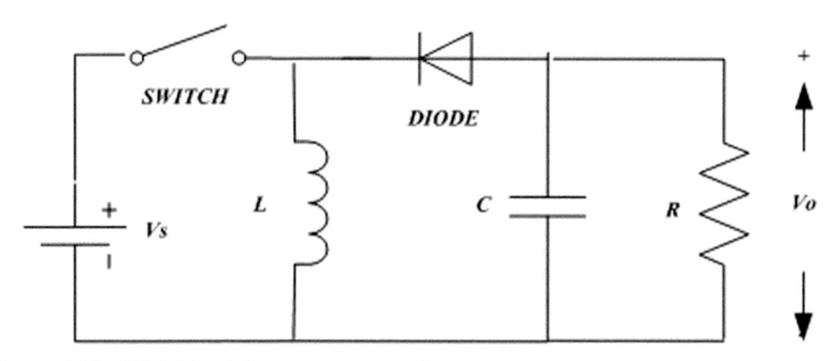

Figure 11.3 DC-DC buck-boost converter circuit.

$$V_O = \frac{-DV_s}{(1-D)} \tag{11.5}$$

$$D = \frac{t_{on}}{t_{on} + t_{off}} \tag{11.6}$$

Table 11.2 Parameters of the buck-boost converter.

Parameters	Symbols	Values
Switching frequency	f	10kHz
Inductor	L	47μH
Capacitor	C	1000μF
Resistive load	R_L	150 Ohm

11.3 Maximum power point tracking system

As it is shown in Fig. 11.2(C), the P-V characteristics carry a single peak power point under uniform atmospheric conditions. When the load is directly connected to the panel, it is rare to constantly drive the PV source at the peak power point. Nevertheless, as the load changes, the operating point also moves from its original position and decreases the efficiency of the PV system. Thus, a controllable power interface between load and the PV source is required. This power interface increases the power transfer capacity of the PV system by adjusting the operating point near PPT. In most cases, DC-DC converters and/or DC-AC converters are used as a power interface in PV system [27]. These converters adjust the operating voltage of the PV source according to the control signal generated by the MPPT system.

The main role of the MPPT system is to find peak power point and provide an appropriate duty ratio for the converters. The most popular and widely used MPPT algorithms are P&O and incremental conductance. The implementations of these algorithms are simple and require low-cost controllers [28]. These algorithms are more suitable under uniform atmospheric conditions and do not need any information about the numbers of PV panels connected and other manufacturing details. However, these algorithms require more tracking time and have continuous power oscillations around MPP. Fuzzy logic-MPPT (FL-MPPT)-based algorithm is introduced to elevate the drawbacks of conventional MPPT algorithms [29]. In most of the FL-MPPT algorithms, the slope of P-V characteristics is used to generate the control signal. This algorithm improves the transient and steady-state tracking performance under all atmospheric conditions. The main drawback of FL-MPPT algorithms is that the operating point moves away from MPP under dynamic atmospheric conditions. In the next section, P&O-based MPPT algorithm and FLC-based MPPT algorithms are discussed in detail and comparative analysis is presented by implementing with the PV system under dynamic atmospheric conditions.

11.3.1 Perturb and observe based maximum power point tracking system algorithm

This algorithm is one of the least complex algorithms, which has been widely used due to its low implementation cost [30]. In this algorithm, the PV voltage is perturbed and the change in its calculated output power is observed; if the calculated power is more than the previous power, again PV voltage is perturbed in the same direction until power becomes less than the previous power; if the value power decreases on increasing the PV voltage, the perturbation of PV voltage is conducted in the opposite direction. At each cycle, such forward and reversed perturbation of PV voltage is recapitulated until they reach the MPP. As it is shown in Fig. 11.4, the MPP is attained when the variation in power becomes zero for every perturbation. The same strategy can be described from the Eq. (11.7a,b,c).

$$V_{pv}(n) = \begin{cases} V_{pv}(n-1) + \Delta V, \text{for} \dfrac{dP_{pv}(n)}{dV_{pv}(n)} > 0 \\ V_{pv}(n-1) + 0, \text{for} \dfrac{dP_{pv}(n)}{dV_{pv}(n)} = 0 \\ V_{pv}(n-1) - \Delta V, \text{for} \dfrac{dP_{pv}(n)}{dV_{pv}(n)} < 0 \end{cases} \tag{11.7a,b,c}$$

Where, "n" represents "nth" sample.

The implementation of the algorithm takes place according to the flowchart illustrated in Fig. 11.5. The figure shows that the flowchart

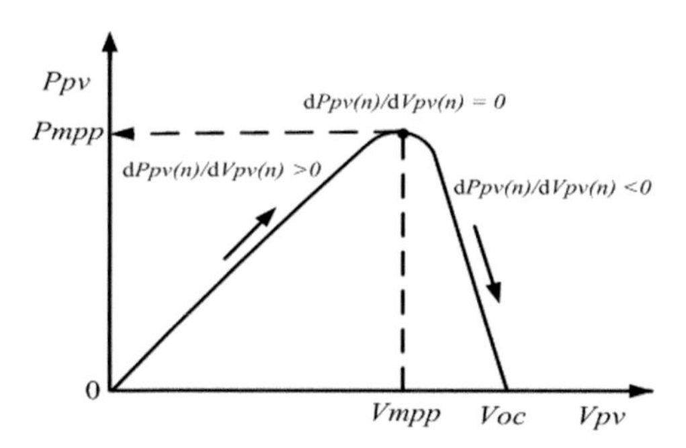

Figure 11.4 P-V characteristics of photovoltaic cell.

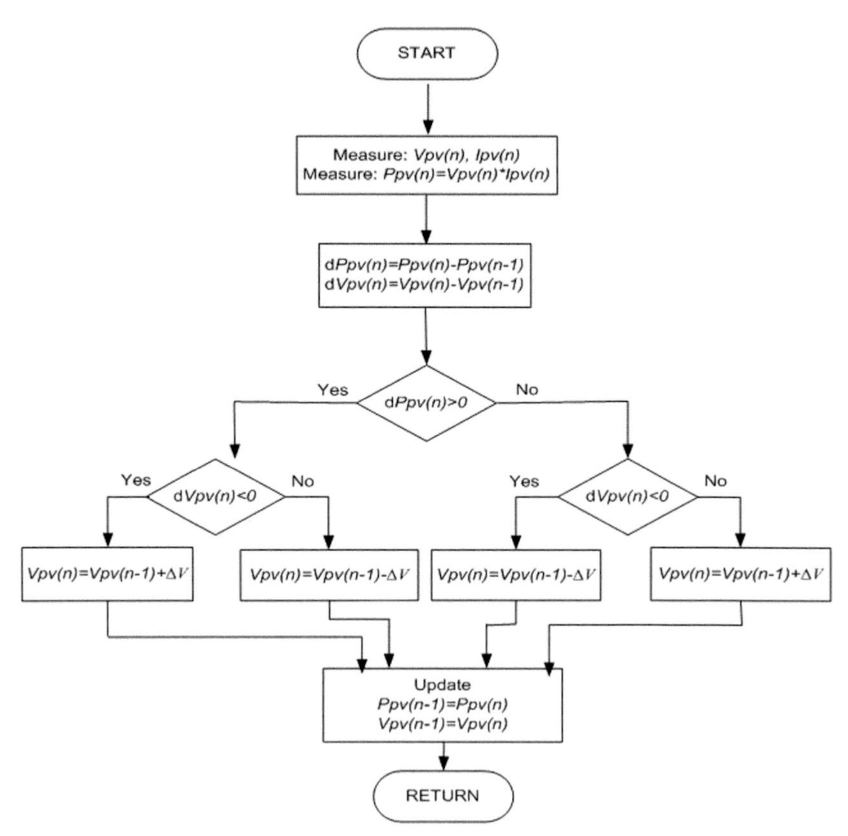

Figure 11.5 Flowchart of perturb and observe based maximum power point tracking algorithm [30].

begins by observing the current PV voltage V_{pv} (n), PV current I_{pv} (n), and PV power P_{pv} (n). At each MPPT cycle, the algorithm perturbs the operating voltage V_{pv} (n) by ΔV, and the PV power P_{pv} (n) is noticed. The working rules of P&O based MPPT algorithm are shown as follows, where $\mathrm{d}V_{pv}(n)$ is the change of voltage and $\mathrm{d}(n)$ is the change of output power as shown in Fig. 8.5.

1. If $\mathrm{d}P_{pv}(n) > 0$ and $\mathrm{d}V_{pv}(n) > 0, \frac{\mathrm{d}P_{pv}(n)}{\mathrm{d}V_{pv}(n)} > 0$, Vpv (n) is increased by ΔV.
2. If $\mathrm{d}P_{pv}(n) > 0$ and $\mathrm{d}V_{pv}(n) > 0, \frac{\mathrm{d}P_{pv}(n)}{\mathrm{d}V_{pv}(n)} < 0$, Vpv (n) is decreased by ΔV.
3. If $\mathrm{d}P_{pv}(n) < 0$ and $\mathrm{d}V_{pv}(n) > 0, \frac{\mathrm{d}P_{pv}(n)}{\mathrm{d}V_{pv}(n)} < 0$, Vpv (n) is increased by ΔV.
4. If $\mathrm{d}P_{pv}(n) < 0$ and $\mathrm{d}V_{pv}(n) < 0, \frac{\mathrm{d}P_{pv}(n)}{\mathrm{d}V_{pv}(n)} > 0$, Vpv (n) is decreased by ΔV.

The above rules are repeated until MPP is achieved. Finally, this algorithm successfully achieves MPP and provides the ideal operating voltage V_{mpp}. The characteristic trait of this algorithm is that it finds MPP without any prerequisite information about atmospheric conditions and solar cell quantities. This algorithm shows better tracking performance under uniform atmospheric conditions.

11.3.2 Design of fuzzy logic based maximum power point tracking system

The composition of FL-MPPT utilizes the user experience about the system behavior. As it is given in Fig. 11.6, FLC has three basic stages: fuzzification stage, decision stage, defuzzification stage [13]. In the fuzzification stage, crisp inputs variables are converted to linguistic variables with the help of predefined MFs. These linguistic variables are utilized to make decisions according to the rules set up by the designer. In the third stage, the generated outputs are converted to crisp variables. For FL-MPPT applications, two inputs are utilized to generate one output in order to reach MPP. The input crisp variables can be described as follows:

$$\frac{dP_{pv}}{dV_{pv}} = E_{pv}(n) = \frac{P_{pv}(n) - P_{pv}(n-1)}{V_{pv}(n) - V_{pv}(n-1)} \tag{11.8}$$

$$\Delta E_{pv}(n) = E_{pv}(n) - E_{pv}(n-1) \tag{11.9}$$

Where E_{pv} (n) is the slope of P-V characteristics for n^{th}, $\Delta E_{pv}(n)$ is the change in E_{pv} (n) between the samples n^{th} and $(n^{th}$-1). Figs. 11.7 and 11.8

First Input ($E_{pv}(n)$): This is the first fuzzy logic controller input. Its negative and positive deviations are measured at STC and classified into seven MFs: NMF_3, NMF_2, NMF_1, ZMF, and PMF_1, PMF_2, and PMF_3 as depicted in Fig. 11.9(A). Where α_1, α_2,... to α_9 are the supports of the MFs.

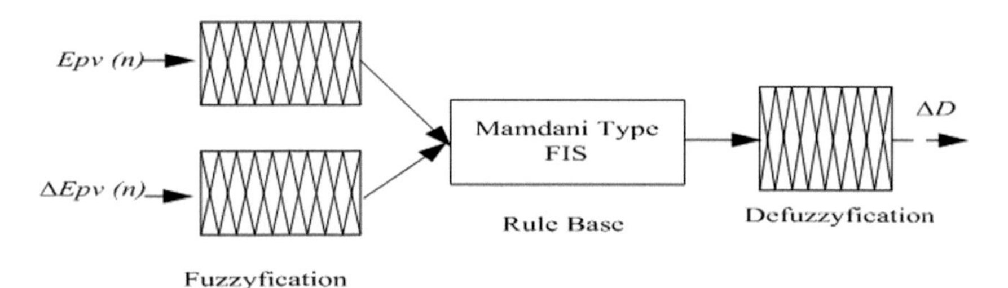

Figure 11.6 Block diagram of fuzzy logic controller [13].

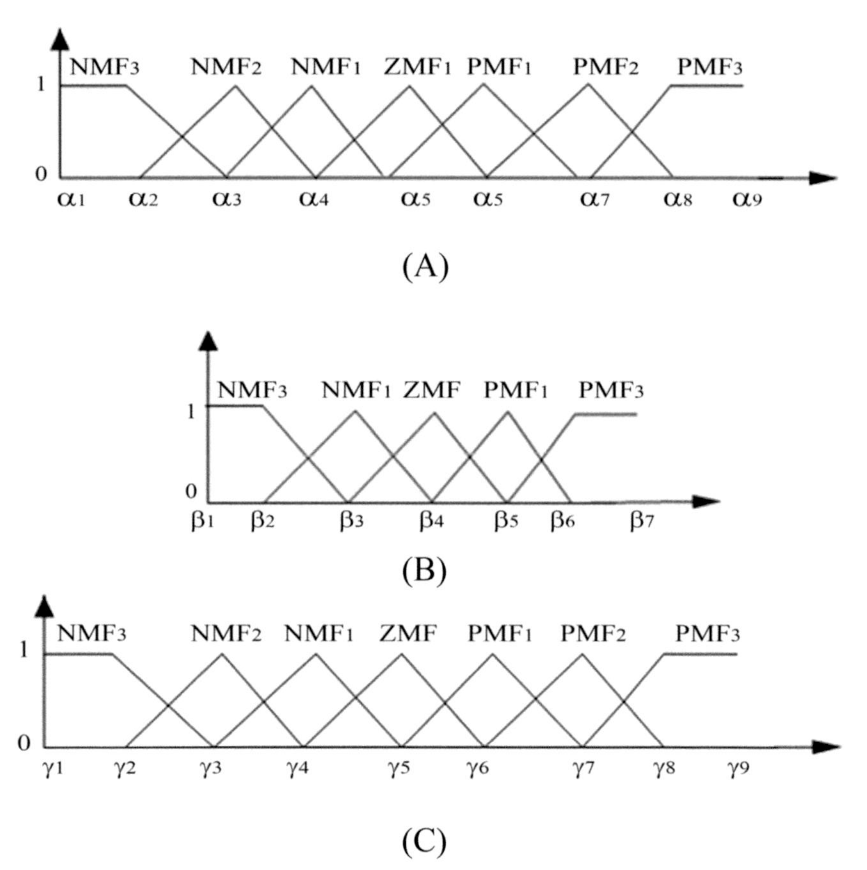

Figure 11.7 Input and output membership functions. (A) MFs of first Input $E_{pv}(n)$, (B) MFs of second input $\Delta E_s(n)$. (C) MFs of output change in duty ratio ΔD.

Second Input (ΔE_{pv}(n)): This is second fuzzy logic controller input. Its negative and positive values are measured at STC and classified into five MFs: NMF_3, NMF_1, ZMF, and PMF_1, and PMF_3s as illustrated in Fig. 11.9(B). Where β_1, β_2...to β_7 are the supports of the MFs. Tables 11.3 and 11.4.

Output (ΔD): Defuzzification is the final stage of the FL controller. Here, the defuzzified output ΔD is the fuzzy controller output which is a variable step size control strategy. Therefore, the output ΔD is classified into seven MFs: NMF_3, NMF_2, NMF_1, ZMF, and PMF_1, PMF_2, and PMF_3 as shown in Fig. 8.7(C). In this system "center of gravity" method is used for the defuzzification of the FIS output. Where, γ_1, γ_2...to γ_9

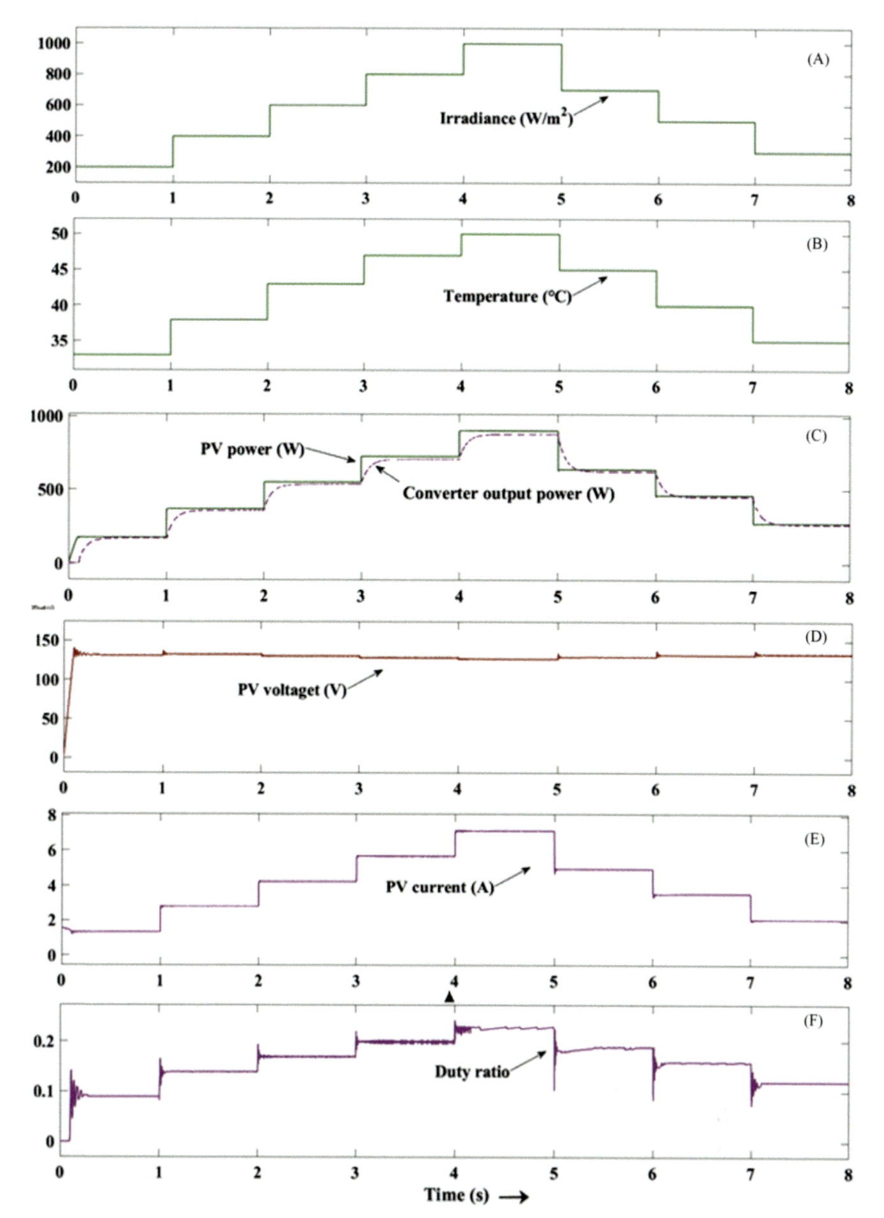

Figure 11.8 Simulation results of fuzzy logic-maximum power point tracking based photovoltaic system under gradual changing atmospheric conditions. (A) Incident irradiance, (B) Module temperature, (C) photovoltaic power and converter output power, (D) photovoltaic voltage, (E) photovoltaic current photovoltaic, and (F) Duty ratio.

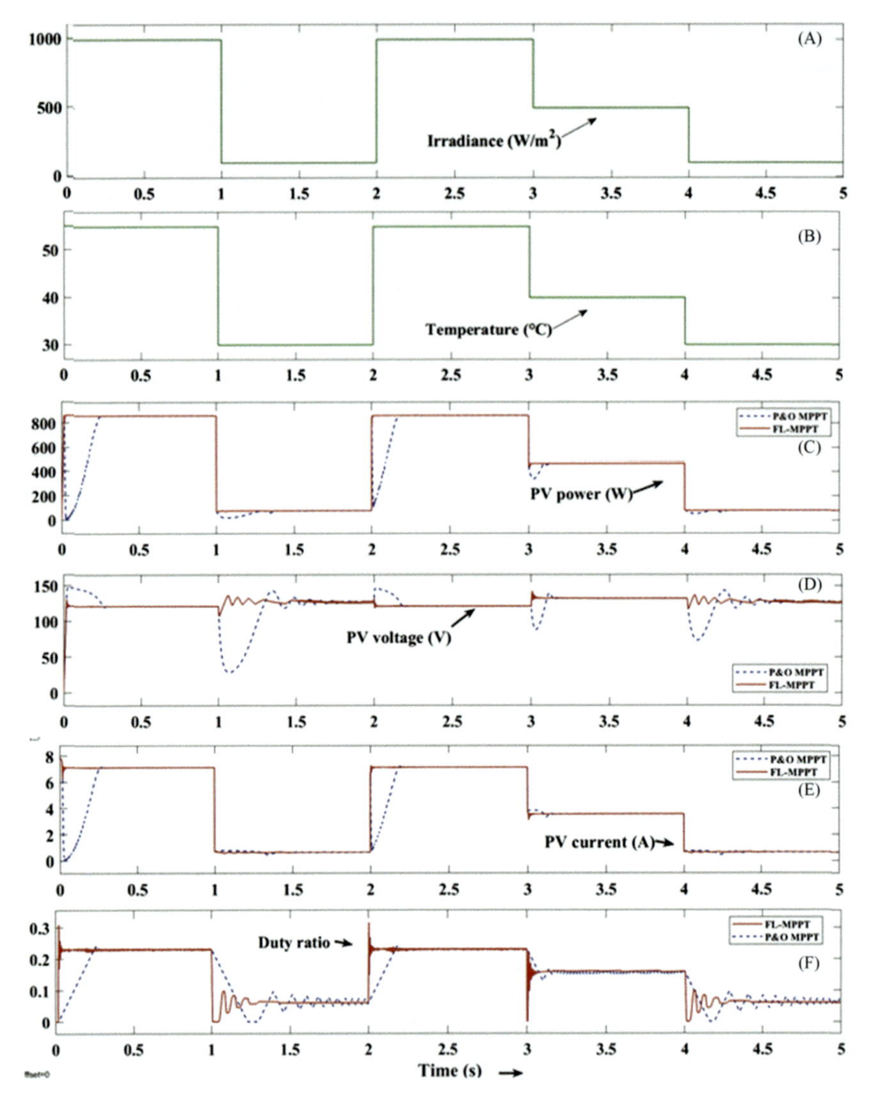

Figure 11.9 Comparative simulation results of fuzzy logic and perturb and observe based photovoltaic-maximum power point tracking algorithms (A) Incident irradiance, (B) Module temperature, (C) photovoltaic module output power, (D) photovoltaic voltage, (E) photovoltaic current and (F) Duty ratio.

are the supports of the MFs. The output of the FLC is then used to calculate the updated duty ratio according to the following equation.

$$D(k) = D(n-1) + \Delta D \tag{11.10}$$

Table 11.3 Nomenclature of membership functions.

NMF_3	Negative big MF
NMF_2	Negative medium MF
NMF_1	Negative short MF
ZMF	Zero MF
PMF_1	Positive short MF
PMF_2	Positive medium MF
PMF_3	Positive big MF

Table 11.4 Complete fuzzy rule base.

$E_{pv}(n)$	$\Delta E_{pv}(n)$				
	NMF_3	NMF_1	ZMF	PMF_1	PMF_3
PMF_3	ZMF	NMF_1	NMF_2	NMF_3	NMF_3
PMF_2	PMF_1	ZMF	NMF_1	NMF_2	NMF_3
PMF_1	PMF_2	PMF_1	ZMF	NMF_1	NMF_3
ZMF	PMF_3	PMF_2	PMF_1	ZMF	NMF_3
NMF_1	PMF_3	PMF_3	PMF_2	PMF_1	NMF_2
NMF_2	PMF_3	PMF_3	PMF_3	PMF_2	NMF_1
NMF_3	PMF_3	PMF_3	PMF_3	PMF_3	ZMF

Where $D(n)$ is the duty ratio of "n^{th}" sample $D(n\text{-}1)$ is the duty ratio of $(n^{th}\text{-}1)$ sample. This updated duty ratio is fed to the pulse width generator to produce a gating signal employed for the converter.

11.4 Simulation results and discussion

The performance analysis of the fuzzy logic-based PV-MPPT algorithm is examined on MATLAB software. The specification of parameters utilized in the simulation work is given in Tables 11.1 and 11.2. The simulation of a PV system is carried out for 1.1 kW of capacity as depicted in the figures. In the first case, the tracking performance of the FL-MPPT algorithm is illustrated in Fig. 11.8. However, in Fig. 11.9, it is compared with P&O based PV-MPPT algorithm to prove its effectiveness under step-changing atmospheric conditions. Finally, the analysis of the results is then summarized in the table for the selection guide.

Fig. 11.8 shows the tracking performance of the FL-MPPT algorithm under dynamic atmospheric conditions. Initially, the irradiance and temperature are increased by 200 W/m^2 and 5°C, respectively, as shown in

Fig. 11.8(A) and (B). During these conditions, the MPP is achieved successfully which can be verified by the PV power and converter output power as shown in Fig. 11.8(C). It can also be proved by the PV voltage, PV current, and the duty ratio values as illustrated in Fig. 11.8(D), (E), and (F), respectively. The same pattern is observed when the values of atmospheric conditions decrease gradually. Therefore, this simulation proves the tracking performance of FL-MPPT under gradual increasing and decreasing atmospheric conditions. Moreover, it is clear from the Fig. 11.8(C), during increasing irradiance conditions, the converter's transient output power is less than the PV power, and during decreasing irradiance conditions it is more than the PV power. It proves the proper working of the converter and the system during the whole simulation.

The FL-MPPT algorithm is a well-known variable step-changing control strategy unlike P&O based PV-MPPT algorithm. This is examined by the PV system simulation under step-changing atmospheric conditions for 5 seconds as shown in Fig. 11.9(A) and (B). Initially, incident irradiance and temperature are kept at 1000 W/m^2 and 50°C for 1 second and then reduced to the values of 100 W/m^2 and 30°C, respectively. As it is shown in Fig. 11.9(C), the peak power tracking speed of FL-MPPT is much faster than the P&O algorithm. The same can be observed from the PV voltages, PV currents, and duty ratios as shown in Fig. 11.9(D), (E), and (F). In the next case, a step increase in the atmospheric condition is applied to the PV system. Here, irradiance and temperature are again increased from 100 W/m^2 and 30°C to 1000 W/m^2 and 50°C, respectively. During this atmospheric condition, the MPP tracking performance of FL-MPPT is again observed to be faster than the P&O algorithm. On stepping the simulation ahead from 3 to 5 seconds, similar comparative results are noted. Thus it is clear that the tracking speed and average

Table 11.5 Performance comparison of fuzzy logic-maximum power point tracking and perturb and observe based photovoltaic-maximum power point tracking.

Comparison parameters	FL-MPPT	P&O
MPP tracking speed	High	Low
MPP tracking accuracy	High	High
Sensors required	2	2
Digital or analog	Digital	Analog/digital
Control strategy	Variable step size	Fixed step size
Complexity	High	Low
Implementation cost	High	Low

power content are high in the case of FL-MPPT as compared to the P&O algorithm under all atmospheric situations.

On the basis of the above results and comparison outcome, a selection guide is presented in the Table 11.5. In this table, the tracking proficiency and control strategies of FL-MPPT and P&O algorithms are compared.

11.5 Conclusion

In this chapter, Fuzzy logic-based PV-MPPT and P&O PV-MPPT algorithms are implemented with a 1.1 kW PV system for performance evaluation and comparison. The tracking performance of FL-MPPT is found comparatively better than the P&O PV-MPPT algorithm. Due to the step-changing control strategy of FL-MPPT, it has a high convergence speed for all varying atmospheric conditions. However, the control complexity is high in the case of FL-MPPT. Therefore, for optimizing the PV system, it is required to make a trade-off between accuracy and complexity.

References

[1] C. Bhattacharjee, B.K. Roy, Fuzzy-supervisory control of a hybrid system to improve contractual grid support with fuzzy proportional-derivative and integral control for power quality improvement, IET Gener. Transm. Distrib. 12 (7) (2018) 1455–1465.
[2] J.R. Nayak, B. Shaw, B. Kumar Sahu, Hybrid alopex based DECRPSO algorithm optimized fuzzy-PID controller for AGC, J. Eng. Res. 8 (1) (2020) 248–271.
[3] P.C. Cheng, B.R. Peng, Y.H. Liu, Y.S. Cheng, J.W. Huang, Optimization of a fuzzy-logic-control-based MPPT algorithm using the particle swarm optimization technique, Energies 8 (6) (2015) 5338–5360.
[4] A.A.S. Mohamed, A. Berzoy, O.A. Mohammed, Design and hardware implementation of FL-MPPT control of PV systems based on GA and small-signal analysis, IEEE Trans. Sustain. Energy 8 (1) (2017) 279–290.
[5] P. Bayat, A. Baghramian, A novel self-tuning type-2 fuzzy maximum power point tracking technique for efficiency enhancement of fuel cell based battery chargers, Int. J. Hydrog. Energy 45 (43) (2020) 23275–23293.
[6] K.Y. Yap, C.R. Sarimuthu, J.M.Y. Lim, Artificial intelligence based MPPT techniques for solar power system: a review, J. Mod. Power Syst. Clean. Energy 8 (6) (2020) 1043–1059.
[7] A. Youssef, M. El Telbany, A. Zekry, Reconfigurable generic FPGA implementation of fuzzy logic controller for MPPT of PV systems, Renew. Sustain. Energy Rev. 82 (2018) 1313–1319.
[8] Y.Y. Hong, P.M.P. Buay, Robust design of type-2 fuzzy logic-based maximum power point tracking for photovoltaics, Sustain. Energy Technol. Assess. 38 (2020) 100669.

[9] T. Ma, H. Yang, L. Lu, Solar photovoltaic system modeling and performance prediction, Renew. Sustain. Energy Rev. 36 (2014) 304–315.
[10] Y. Chaibi, A. Allouhi, M. Salhi, A. El-jouni, Annual performance analysis of different maximum power point tracking techniques used in photovoltaic systems, Prot. Control Mod. Power Syst. 4 (1) (2019) 1–10.
[11] X. Li, Q. Wang, H. Wen, W. Xiao, Comprehensive studies on operational principles for maximum power point tracking in photovoltaic systems, IEEE Access 7 (2019) 121407–121420.
[12] B. Subudhi, R. Pradhan, A comparative study on maximum power point tracking techniques for photovoltaic power systems, IEEE Trans. Sustain. Energy 4 (1) (2013) 89–98.
[13] A.G. Al-Gizi, S.J. Al-Chlaihawi, Study of FLC based MPPT in comparison with P&O and InC for PV systems, in: 2016 International Symposium on Fundamentals of Electrical Engineering (ISFEE), 2016, pp. 1–6, doi: 10.1109/ISFEE.2016.7803187.
[14] R. Sahu, Design of solar system by implementing ALO optimized PID based MPPT controller, Trends Renew. Energy 4 (3) (2018) 44–55.
[15] S. Jain, V. Agarwal, A new algorithm for rapid tracking of approximate maximum power point in photovoltaic systems, IEEE Power Electron. Lett. 2 (1) (2004) 16–19.
[16] D. Choudhary, A. Ratna Saxena, Incremental conductance MPPT algorithm for PV system implemented using DC-DC buck and boost converter, Int. J. Eng. Res. Appl. 4 (8) (2014) 123–132. Available from: http://www.ijera.com.
[17] J. Liu, J. Li, J. Wu, W. Zhou, Global MPPT algorithm with coordinated control of PSO and INC for rooftop PV array, J. Eng. 2017 (13) (2017) 778–782. Available from: https://doi.org/10.1049/joe.2017.0437.
[18] S.Z.Muhammad Nur, A.M. Umar, M.A.M. Radzi, N.N. Mahzan, Single stage string inverter for grid- connected photovoltaic system with modified perturb and observe (P&O) fuzzy logic control (FLC)-based MPPT technique, J. Electr. Syst. 12 (2) (2016) 344–356.
[19] A. Bouchakour, A. Borni, M. Brahami, Comparative study of P&O-PI and fuzzy-PI MPPT controllers and their optimisation using GA and PSO for photovoltaic water pumping systems, Int. J. Ambient. Energy 0750 (2019).
[20] P. Singh, S. Bhardwaj, S. Dixit, R.N. Shaw, A. Ghosh, Development of prediction models to determine compressive strength and workability of sustainable concrete with ANN, in: S. Mekhilef, M. Favorskaya, R.K. Pandey, R.N. Shaw (Eds.), Innovations in Electrical and Electronic Engineering. Lecture Notes in Electrical Engineering, vol. 756, Springer, Singapore, 2021. Available from: https://doi.org/10.1007/978-981-16-0749-3_59.
[21] Y. Belkhier, A. Achour, R.N. Shaw, A. Ghosh, Performance improvement for PMSG tidal power conversion system with fuzzy gain supervisor passivity-based current control, in: S. Mekhilef, M. Favorskaya, R.K. Pandey, R.N. Shaw (Eds.), Innovations in Electrical and Electronic Engineering. Lecture Notes in Electrical Engineering, vol. 756, Springer, Singapore, 2021. Available from: https://doi.org/10.1007/978-981-16-0749-3_6.
[22] R.K. Kharb, S.L. Shimi, S. Chatterji, M.F. Ansari, Modeling of solar PV module and maximum power point tracking using ANFIS, Renew. Sustain. Energy Rev. 33 (2014) 602–612.
[23] R.N. Shaw, P. Walde, A. Ghosh, IOT based MPPT for performance improvement of solar PV arrays operating under partial shade dispersion, in: 2020 IEEE 9th Power India International Conference (PIICON), 2020, pp. 1-4, doi: 10.1109/PIICON49524.2020.9112952.

[24] H. Bellia, R. Youcef, M. Fatima, A detailed modeling of photovoltaic module using MATLAB, NRIAG J. Astron. Geophys. 3 (1) (2014) 53–61.
[25] L. An, D.D.C. Lu, Design of a single-switch DC/DC converter for a PV-battery-powered pump system with PFM + PWM control, IEEE Trans. Ind. Electron. 62 (2) (2015) 910–921.
[26] J.W. Zapata, S. Kouro, G. Carrasco, H. Renaudineau, T.A. Meynard, Analysis of partial power DC-DC converters for two-stage photovoltaic systems, IEEE J. Emerg. Sel. Top. Power Electron. 7 (1) (2019) 591–603.
[27] V.C. Kotak, P. Tyagi, DC To DC converter in maximum power point tracker, Int. J. Adv. Res. Electr. Electron. Instrum. Eng. 2 (12) (2007) 6115–6125.
[28] S.A. Mohamed, M. Abd El Sattar, A comparative study of P&O and INC maximum power point tracking techniques for grid-connected PV systems, SN Appl. Sci. 1 (2) (2019) 1–13.
[29] A. Chandran, B.K. Mathew, Comparative analysis of PO and fuzzy logic controller based MPPT in a solar cell, in: Proc. 3rd Int. Conf. Smart Syst. Inven. Technol. ICSSIT 2020, no. Icssit, pp. 622–626, 2020.
[30] A.K. Podder, N.K. Roy, H.R. Pota, MPPT methods for solar PV systems: a critical review based on tracking nature, IET Renew. Power Gener. 13 (10) (2019) 1615–1632.

CHAPTER TWELVE

Control strategy for design and performance evaluation of hybrid renewable energy system using neural network controller

Ashwani Kumar[1], Vishnu Mohan Mishra[2] and Rakesh Ranjan[3]
[1]Department of Electrical Engineering, Uttrakhand Technical University, Sudhowala, India
[2]Department of Electrical Engineering, G. B. Pant Engineering College, New Delhi, India
[3]Department of Electrical Engineering, Himgiri Zee University, Dehradun, India

12.1 Introduction

Over the last few decades mechanical energies such as coal, gas, oil, etc., have been depleting day by day. Such restricted accessibility of these non-renewable energy sources leads to increase in usage of renewable energy sources and there is quick development in renewable energy resources. Because of easy accessibility, being recyclable and eco-friendly, it does not discharge any unsafe gases [1]. Out of them PV and wind are the quickest developing energy sources which are acceptable energy sources around the world.

Due to global industrialization and rapid growth in population, demand of continuous electrical energy is increasing [2]. At present conventional power plants are not able to accommodate this increased energy demand. Available renewable energy source can meet the required energy especially in remote area where continuous supply of energy by conventional means is not possible [3]. In this chapter we are addressing hybrid power system model consisting of PV and wind energy power plant. PV energy model consists of modules and MPPT, utilizing perturb and observe algorithm that extract maximum power from plant [4].

The hybrid energy system considered here is depicted in the Fig. 12.1. The fundamental parts of WECS have wind turbine (WT), gearbox, rotor, generator and converter [5]. Energy storage device like battery bank may be connected to DC terminals of bidirectional converter which

Applications of AI and IOT in Renewable Energy.
DOI: https://doi.org/10.1016/B978-0-323-91699-8.00012-7

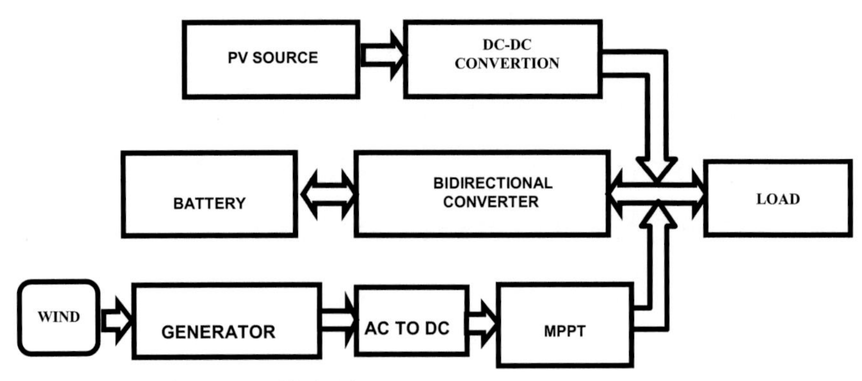

Figure 12.1 Architecture of hybrid system.

store energy generated by solar system and constant power delivered by WT. Such hybrid configuration ensures efficient and reliable supply as compare to single power source system [6].

Optimization performs leading role for the enhancement of control performance of hybrid renewable energy system [6]. A tuning algorithm is a process of getting the best solution of objective function obtained by regulation of the control parameter and comparing various solutions for PID control for the power system by artificial intelligence like ANN [7]. Artificial neural network is one of the soft computing algorithms which are encouraged by biological behavior of living beings [8]. Therefore a hybrid power system is proposed in this chapter with PID controllers tuned by ANN [9]. The system considered here can reduce the fuel consumption in diesel generator system and is proven to be environment friendly.

12.2 Modeling of hybrid power system

Load sharing is one of the methods of load management by distributing the load between available energy generating units operating in parallel [10]. Here load distribution is done between parallel operating generator sets in proportional to the KW and KVAR. For the purpose of load sharing dropping control strategy is widely used [11]. With this technique, generators are involved to regulate and sharing of load in equal or fractional part of rated capacity of generator as per need of frequency and voltage [12]. This suggested control strategy is used to control primary frequency of generator applicable to parallel operation.

In parallel operation of generators, the idea of drooping control is used, in which there is decrement in frequency with the increment in load due system inertia [13]. In Micro grids, all the sources having almost zero inertia are connected through power electronics interfacing. Due to this phenomenon, conventional turbine generator system reproduces the inertia in VSI's using drooping characteristics of power controller devices [14]. The voltage drop characteristics are shown in Fig. 12.2. It is also defined as P-W & E-Q characteristics.

From inverter to grid through impedance P&Q power flow can be represented as

$$P = \left(\frac{\mathrm{EV}}{Z}cos(\delta) - \frac{V^2}{Z}\right)cos(\theta) + \frac{\mathrm{EV}}{Z}\sin(\theta)\sin(\delta) \tag{12.1}$$

$$\mathrm{Q} = \left(\frac{\mathrm{EV}}{Z}cos(\delta) - \frac{V^2}{Z}\right)sin(\theta) - \frac{\mathrm{EV}}{Z}\cos(\theta)\sin(\delta) \tag{12.2}$$

Here Z and θ represents the output impedance & phase difference, respectively. δ and V depicts the phase angle and common bus voltage, respectively [15]. Without decoupling between *P-W* and *E-Q* characteristics values of $\theta = 90$ & $Z = X$, is considered. Hence active and reactive power can be shown as:

$$P = \frac{\mathrm{EV}}{X}\sin(\delta) \tag{12.3}$$

$$\mathrm{Q} = \frac{\mathrm{EV}}{X}\cos(\delta) - \frac{V^2}{X} \tag{12.4}$$

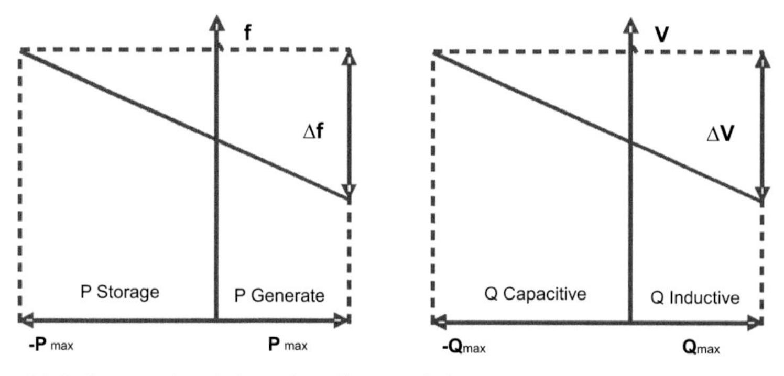

Figure 12.2 Conventional drooping characteristics.

With negligible values of δ, $\sin\delta = \delta$ & $\cos\delta = 1$

$$P = \frac{EV}{X}\delta \tag{12.5}$$

$$Q = \left(\frac{EV}{X} - \frac{V^2}{X}\right) \tag{12.6}$$

From above equation it is inferred that linear relationship between P-W and E-Q [16].

$$\omega = \omega^* - m_f P \tag{12.7}$$

$$E = E^* - m_f Q \tag{12.8}$$

where ω is the inverter frequency, E represents the magnitude of inverter output voltage. In the same way, at no-load the frequency and magnitude of inverter output voltage is defined and shows the proportional droop coefficient of voltage and frequency [17].

In our model, load is shared between PV, Wind (PMSG) and Battery storage based power plants. Here, load is applied for higher capacity then of WT system such that first wind is fully utilized and later additional require amount of power is supplied by PV system. Artificial drooping concept is introduced to control the load sharing between both systems. The model contains WT containing permanent synchronous machine with round type rotor with rotating speed 3000 rpm and a solar thermal power plant as source of energy. Radial Basic Function Neural Network is used for tuning purpose. PID controller for controlling the Power Sources. The modeling and simulation of purposed model is performed on MATLAB SIMULINK.

12.3 Control strategy

The strategy is implemented by regulating the error in power which can be shown as:

$$\Delta P = P_D - P_G \tag{12.9}$$

The power error which is the difference between the load demand P_D and net power generated P_G. Variation in generating power leads to change in frequency response in hybrid power system [18]. This can be represented by:

$$\Delta F = \frac{\Delta P}{k} \tag{12.10}$$

Table 12.1 Parameters of the hybrid system.

Sr. no.	Source	Gain	Time constant (s)
1	Solar thermal	4	0.4
2	Wind	0.2	0.2
3	Battery storage system	6	0.16

Where ΔP and k shows the change in power generation and is system frequency, respectively. From above, equation may be:

$$\frac{\Delta F}{\Delta P} = \frac{1}{k(1 + T_s)} \tag{12.11}$$

Eq. (12.9) can also be represented as:

$$\Delta F = \frac{1}{D + sM} \Delta P \tag{12.12}$$

Here $M = 0.012$ and $D = 0.2$ depicts the inertia constant & damping constant. Different parameters considered here are shown in Table 12.1.

12.3.1 Neural network model

In this attempt, a radial basis function is implemented as an activation function model of multilayer feed forward ANN model with three layers. In which 1st layer is linear which distributes the input signal while the incoming layer is nonlinear. The third layer performs the function of combination of outputs. Here the modification in weights done between the hidden layer and the output layer. Normally radial basis function have different parameters for optimization, like the weight between the hidden layer and the output layer, activation link, center of activation function, distribution of center of activation function and count of hidden neurons. It is acceptable for real time application due to quick training duration and general approach. Proper selection of radial basis function is selected by Gaussian kernel for PID tuning application.

Fig. 12.3 shows the ANN model just as implemented in MATLAB's RBF Toolbox. The suggested ANN radial basis function model is implemented with unknown number of inputs and output layer values within the PID parameter limits of the power system. Here back propagation method is engaged to train the ANN and regulate its weight in direction to minimize the error between expected and theoretical values at every iteration. MATLAB's RBF Toolbox is used to build and train 1000

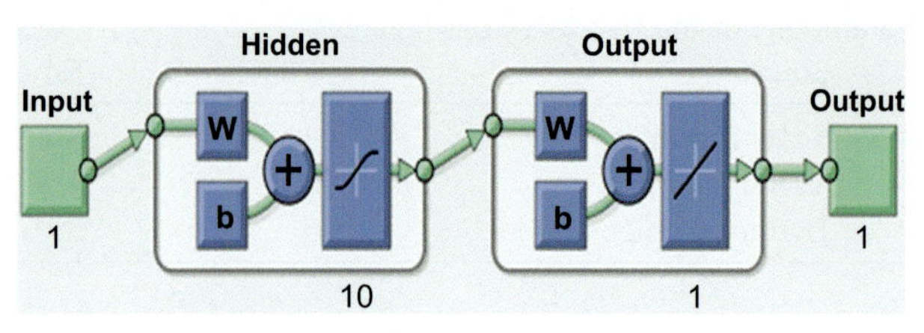

Figure 12.3 Neural network model.

random values both for input and output. PID parameters may represent the input data and ISE shows the output data of load variation from hybrid power disturbances.

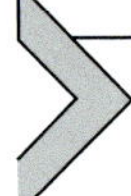

12.4 Proportional-integral-derivative control and performance index

The PID controller is a combination of different control actions which minimizes the error between load demand and power generation. Transfer function for PID controller is shown as:

$$C(s) = K_p + \frac{K_i}{s} + K_d \tag{12.13}$$

The performance optimization of hybrid power system controller can be done by regulating the tuning of performance index. Lower value of performance index is preferred for making the system robust against power fluctuation. Integral square error (ISE) used as objective function to design optimum system. The ISE objective function is calculated as Eq. (12.14), where ΔP and t represent the power fluctuation (pu), the load deviation (pu), and the time period of the simulation respectively.

$$\text{ISE} = \int_0^\infty e^2(t)\text{d}t \tag{12.14}$$

The proposed system model is shown in Fig. 12.3. This model is designed in MATLAB/ Simulink Software.

12.5 Simulation results and discussion

In this section, simulation results of different energy sources and their investigation is explored. The different input power sources that are integrated in the proposed model are the WT, solar thermal power plant, battery storage system, etc. This chapter includes a multilayer feed forward ANN RBF model consisting of one input and output neuron. This is realized to find the tuned parameter of the PID controller. Normally back propagation algorithm is employed for the training and regulates the neural network. It also utilized to reduce the error between expected values and the theoretical values for each iteration. MATLAB radial basis function toolbox is used to built and train the model. For the training of the network 1000 arbitrary values are produced for both for input and output data. Fig. 12.4 shows the ISE during the whole iteration process. At the end of the whole iteration process ISE is found to be minimum. Fig. 12.5 depicts the transient state parameter at the end of 1000 epochs.

Bayesian-regularization tuning algorithm is implemented to train and test the data. Figs. 12.5 and 12.6 shows the training and fitting the network model and which is done before the process of determination of coefficient of regression and least mean square error. Fig. 12.6 shows the error function behavior all along the initial training process. This is a tool used indirectly for the determination of efficiency of neural network.

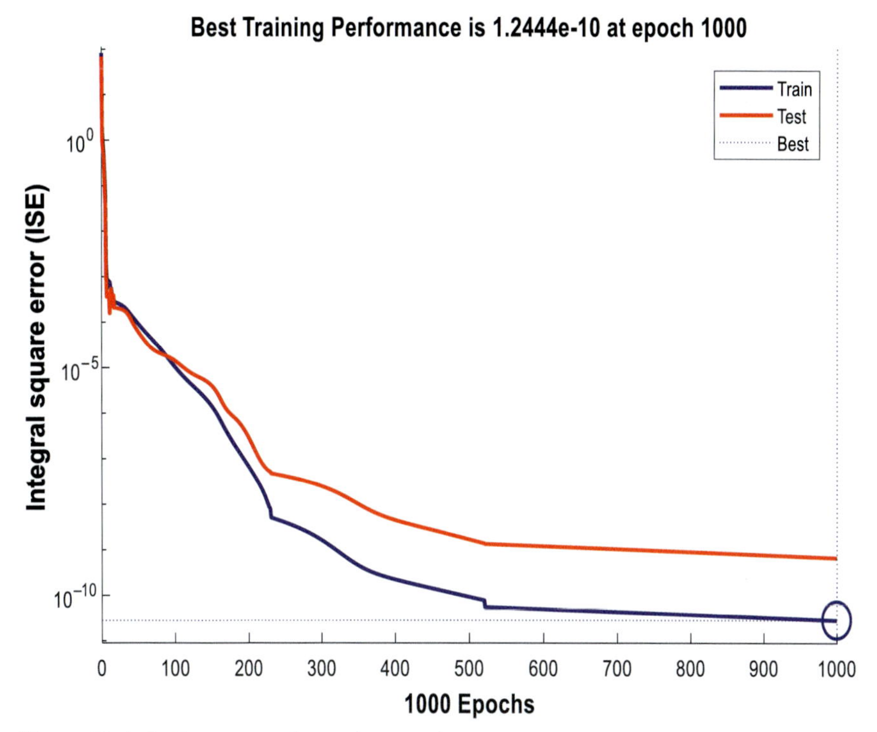

Figure 12.4 Performance of neural network.

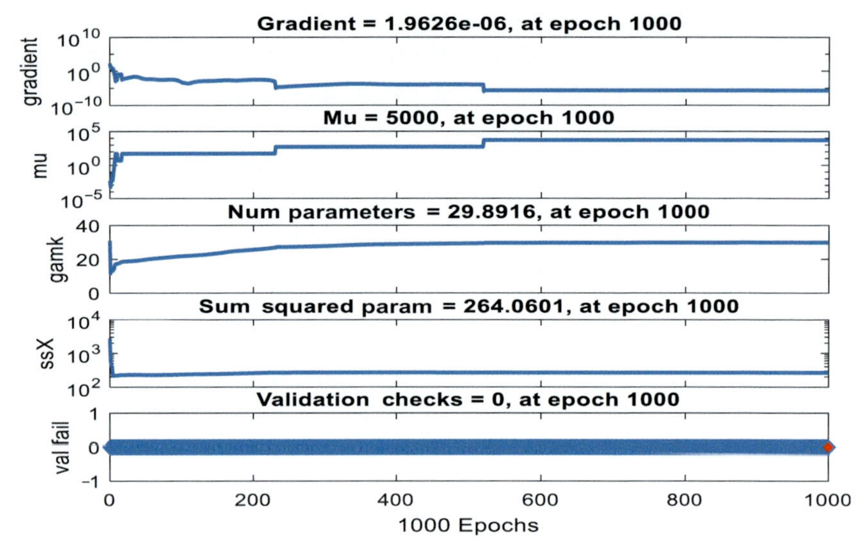

Figure 12.5 Training state parameter at 1000 epochs.

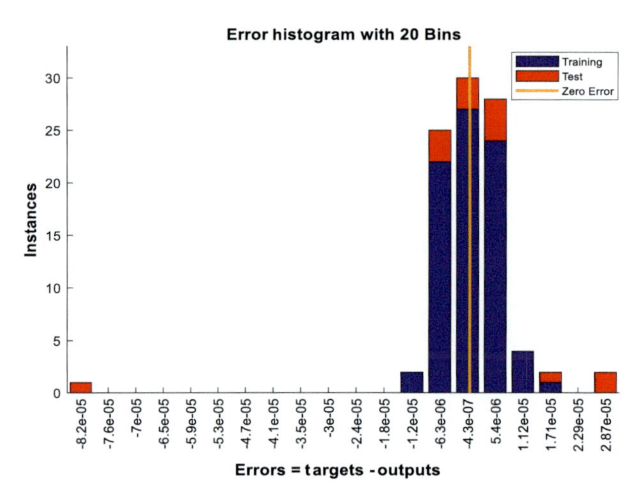

Figure 12.6 Error function behavior.

All the error values of iterations are shown in Fig. 12.7. This shows the iteration taken to tune the neural network. In present work 1000 iterations are taken to optimize the neural. Further error function behavior of the model is shown in Fig. 12.6. Where error is found to be less after neural network training. The regression coefficients to find out the

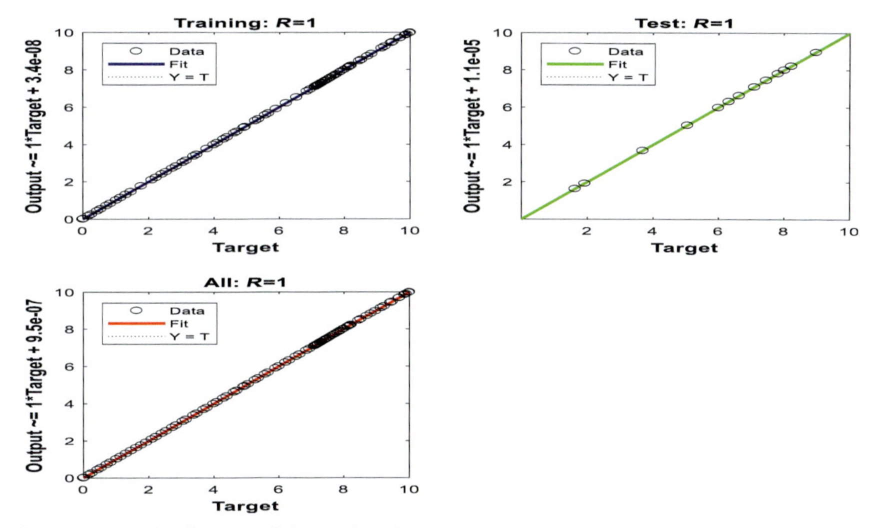

Figure 12.7 Coefficient of determination.

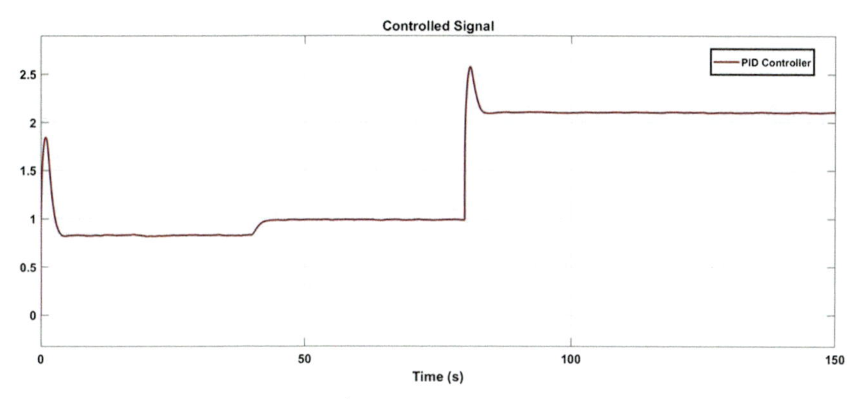

Figure 12.8 Controlled signal performance.

R and IAE post training and fitness to the network model are shown in Fig. 12.7.

The coefficients of determination measure the correlation between real and expected values. It determine the stability of linear relationship among real and expected value that is, $R = 1$. The controlled signal is shown in the Fig. 12.8. The controlling signal is achieved 1.8 and 2.51, respectively.

The waveform shown in Fig. 12.9 represents the input power of wind generator etc. Also Fig. 12.10 shows the feedback power of PID controller, backup power of battery energy storage system and integrated power

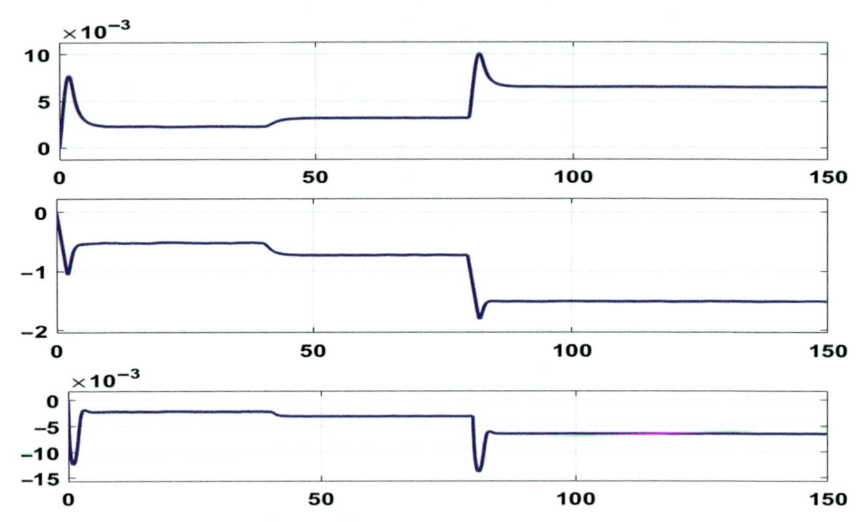

Figure 12.9 Input power from different energy source.

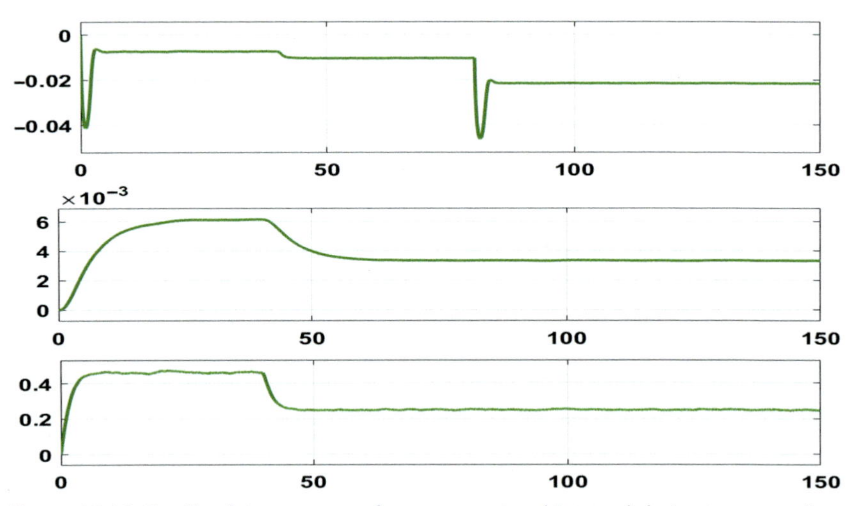

Figure 12.10 Feedback input power from proportional-integral-derivative controller.

of solar cell and WT. The model is simulating for 150 seconds. The output received power from solar and WT is calculated. The output waveform is calculated and represented in Fig. 12.11 which shows the output power of solar thermal and hybrid system. Fig. 12.12 shows wind power output and load demand of WT generator with gain waveform. The

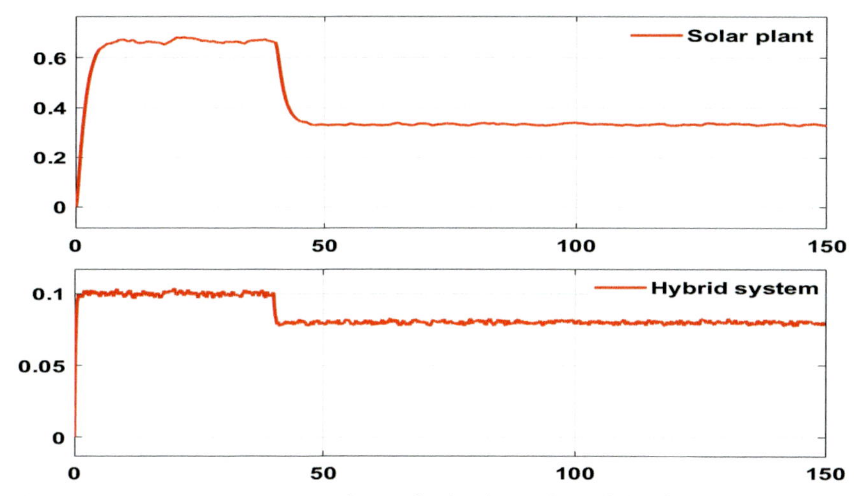

Figure 12.11 Output power waveform of solar thermal and hybrid system.

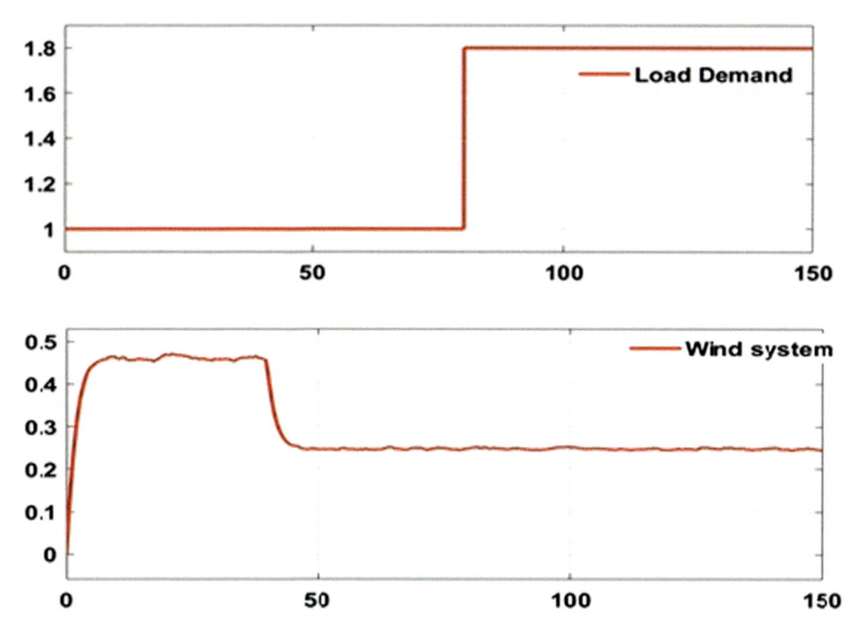

Figure 12.12 Output power waveform of wind source and load demand.

comparative analysis of proposed model & previous model is represented by Table 12.2. The frequency deviation, control signal deviation, and IAE of proposed model are 0.28, 2.23, and 1.05, respectively.

Table 12.2 Comparative analysis.

Sr.no.	Parameter	Proposed approach	Conventional approach
1	Frequency deviation	0.28	0.5
2	Control signal deviation	2.23	2.59
3	ISE	1.05	1.19

12.6 Conclusions

The purpose of suggested work is to preserve the constant flow of power between load and hybrid generating units. This chapter proposes a PID, Neural PID controller for hybrid power system. The neural network is aimed to optimize the PID parameter of controller. Stability is tested against load variation at different instances of time. MATLAB simulation of hybrid power system confirms the effectiveness of neural PID performance. Results clearly show that ISE as an objective function gives less value for frequency deviation in neural PID. The frequency deviation of neural PID controller for suggested hybrid system is 0.28. The control signal deviation of neural PID controller for hybrid system is 2.23. The ISE of neural PID controller for hybrid system is 1.05.

References

[1] L.W. Chong, et al., An optimal control strategy for standalone PV system with battery-super-capacitor hybrid energy storage system, J. Power Sources 331 (2016) 553–565. Available from: https://doi.org/10.1016/j.jpowsour.2016.09.061.

[2] A.H.M. Abdur Rahim, E.P. Nowicki, Super-capacitor energy storage system for fault ride-through of a DFIG wind generation system, Energy Convers. Manage. 59 (2012) 96–102. Available from: https://doi.org/10.1016/j.enconman.2012.03.03.

[3] X. Li, D. Hui, X. Lai, Battery energy storage station (BESS)-based smoothing control of photovoltaic (PV) and wind power generation fluctuations, IEEE Trans. Sustain. Energy 4 (2) (2013) 464–473. Available from: https://doi.org/10.1109/TSTE.2013.2247428.

[4] A.F. Obando-Montaño, et al., A STATCOM with super-capacitors for low-voltage ride-through in fixed-speed wind turbines, Energies 7 (9) (2014) 5922–5952. Available from: https://doi.org/10.3390/en7095922.

[5] Shi, J. et al., Fuzzy logic control of DSTATCOM for improving power quality and dynamic performance, in: Proceedings of the Australasian Universities Power Engineering Conference (AUPEC), 2015, pp. 1-6, doi:10.1109/AUPEC.2015.7324796.

[6] J.J. Justo, F. Mwasilu, J.W. Jung, Doubly-fed induction generator based wind turbines: A comprehensive review of fault ride-through strategies, Renew. Sustain. Energy Rev. 45 (2015) 447–467. Available from: https://doi.org/10.1016/j.rser.2015.01.064.

[7] H. Heydari-Doostabad, M.R. Khalghani, M.H. Khooban, A novel control system design to improve LVRT capability of fixed speed wind turbines using STATCOM in presence of voltage fault, Int. J. Electr. Power Energy Syst. 77 (2016) 280–286. Available from: https://doi.org/10.1016/j.ijepes.2015.11.011.

[8] L. Yang, et al., Coordinated-control strategy of photovoltaic converters and static synchronous compensators for power system fault ride-through, Electr. Power Compon. Syst. 44 (15) (2016) 1683−1692. Available from: https://doi.org/10.1080/15325008.2016.1194502.
[9] Bhangale, S.S., Patel, N., Design of LVRT capability for grid connected PV system, in: Proceedings of the International Conference on Intelligent Computing, Instrumentation and Control Technologies, ICICICT 2017, 2018, pp. 1625−1630. doi: 10.1109/ICICICT1.2017.8342814.
[10] G. Kapoor, V.K. Mishra, R.N. Shaw, A. Ghosh (2021) Fault detection in power transmission system using reverse biorthogonal wavelet, in: Mekhilef S., Favorskaya M., Pandey R.K., Shaw R.N. (Eds.), Innovations in Electrical and Electronic Engineering. Lecture Notes in Electrical Engineering, vol. 756. Springer, Singapore, pp. 381-393. <https://doi.org/10.1007/978-981-16-0749-3_28>.
[11] B. Manikanta, G. Kesavarao, S. Talati, LVRT of Grid Connected PV System with Energy Storage, International Science Press, 2017.
[12] Y. Belkhier, A. Achour, R.N. Shaw, W. Sahraoui, A. Ghosh (2021). Adaptive linear feedback energy-based backstepping and PID control strategy for PMSG driven by a grid-connected wind turbine. In: Mekhilef S., Favorskaya M., Pandey R.K., Shaw R.N. (Eds.), Innovations in Electrical and Electronic Engineering. Lecture Notes in Electrical Engineering, vol. 756. Springer, Singapore, pp. 177-189. <https://doi.org/10.1007/978-981-16-0749-3_13>.
[13] T. Praveen Kumar, N. Subrahmanyam, M. Sydulu (2019). Power flow management of the grid-connected hybrid renewable energy system: a PLSANN control approach, IETE J. Res., 67(4), 569−584. doi:10.1080/03772063.2019.1565950.
[14] P. Singh, S. Bhardwaj, S. Dixit, R.N. Shaw, A. Ghosh (2021). Development of prediction models to determine compressive strength and workability of sustainable concrete with ANN, in: Mekhilef S., Favorskaya M., Pandey R.K., Shaw R.N. (Eds.), Innovations in Electrical and Electronic Engineering. Lecture Notes in Electrical Engineering, vol. 756. Springer, Singapore, pp. 753-769. <https://doi.org/10.1007/978-981-16-0749-3_59>.
[15] T.A. Rashid, D.K. Abbas, Y.K. Turel, A multi hidden recurrent neural network with a modified grey wolf optimizer, PLOS One 14 (3) (2019) e0213237. Available from: https://doi.org/10.1371/journal.pone.0213237.
[16] Y. Belkhier, A. Achour, R.N. Shaw, A. Ghosh (2021) Performance improvement for PMSG tidal power conversion system with fuzzy gain supervisor passivity-based current control, in: Mekhilef S., Favorskaya M., Pandey R.K., Shaw R.N. (Eds.), Innovations in Electrical and Electronic Engineering. Lecture Notes in Electrical Engineering, vol. 756. Springer, Singapore, pp. 81-93. <https://doi.org/10.1007/978-981-16-0749-3_6>.
[17] R. Alireza, E. Ali, E. Hasan, Intelligent hybrid power generation system using new hybrid fuzzy-neural for photovoltaic system and RBFNSM for wind turbine in the grid connected mode, Front. Energy 13 (2019) 131−148.
[18] D.K. Lal, A.K. Barisal, Combined load frequency and terminal voltage control of power systems using moth flame optimization algorithm, J. Electr. Syst. Inf. Technol. 6 (8) (2019) 1−24.

Index

Note: Page numbers followed by "*f*" and "*t*" refer to figures and tables, respectively.

S

T

U

V

W

Z

Printed in the United States
by Baker & Taylor Publisher Services